CW01335657

Shambles

by
Stephen Tait

Jane
Thanks for taking the
time to read 'Shambles'
Stephen Tait

Eloquent Books
New York, New York

Strategic Book Publishing
An imprint of AEG Publishing Group
845 Third Avenue, 6th Floor – 6016
New York, NY 10022
www.eloquentbooks.com

ISBN: 978-1-60860-044-1   1-60860-044-1

Printed in the United States of America

Book Design: Arlinda Van, Dedicated Business Solutions, Inc.

This is a true story. Nearly all the names have been changed. It seemed fair to do so.

CONTENTS

1

Aftermath

One Sunday, late in 2003, following years of indecision as to whether I should or could make this journey, I was standing in a churchyard in Yorkshire, England. A fine rain was drizzling down onto my back, and almost as a tribute to past times, the cold, strong wind from the north was forcing its way through every possible opening in my waxed jacket, chilling me to the bone. Standing alone, transfixed by the tranquillity of this place, I thought my thoughts of four young men. Four young men, who had been friends in the hostile environment of Antarctica, battered, almost into submission on the ice-cap, by a blizzard that had held such aggressive power it had shocked even the more experienced of the small traveling party. Two of the young men, Joe Watson and Ron George, had remained in Antarctica, encased in ice and buried deep in the belly of a crevasse, their resting-place for eternity. I had made my journey to this church to recall past times and to make my peace with Joe, a young man who, more than a lifetime before, used to attend this church with the members of his close-knit family.

Mary Magdalene, carved in marble, drew my eyes to her in her domination of the neatly manicured gardens. Her arms were outstretched, beckoning to the unknown and countless people who must have stood in this very place over the years, seeking answers or simply basking in the serenity I was now experiencing. I felt an inner peace now, a peace that I had not felt for as long as time itself. Joe, my long dead friend and traveling companion, was close to me here; I could feel his presence in the silence. I told him that I had understood his fear of those final days, that I had shared his fear and discomfort and, that at last, the years had given me the understanding of why that final blizzard had ravaged his trust in me.

Tumbling disjointedly through my mind, the events of all those years before came to me in bright flashes of white and orange and blue, dominant colors in the dominating continent of Antarctica. I thought of the violent winds we encountered on our journeys battering against our, orange, laughter filled tent. Joe's recently shaven head, his ragged tartan shirt, and his relaxed and smiling approach to life. The crunch and squeak of the hard, brilliantly white snow under the rubber caterpillar tracks of our machines as we willingly explored the limits of our world, with me always in front and Joe following stoically in my tracks. And I thought of the devastatingly hard times we had endured in the days leading up to the accident, and the devastatingly hard times I had had to endure since.

Joe, my traveling companion, who died in that crevasse on the glacier, a remote and unforgiving place, and Ron, the friend who had died there with him, had in turn, been true friends. They had spent much of their time together during a very short period of their lives and now they were a very short distance apart where they were spending their time together in death.

As the drizzle turned into rain in that Yorkshire church-yard, I heard the latch being lifted on the wrought iron gate that was shielding me from the outside world. I felt my peace would now be shattered and this precious moment lost. There was the murmur of muted conversation as the friend who had accompanied me on this long overdue pilgrimage spoke with the newcomer. I glanced across and saw the response was a nodded understanding and an obvious remembrance of those long past events, and the newcomer, the priest, passed me by in a rain and windswept shimmer of black without making an attempt to interrupt my thoughts. Moments later, with the briefest of glances in my direction, the priest stood aside from the opened church door and accepted openly and without question that I could enter into his church and have a few final moments there in undisturbed solitude with my memories.

Driving away from the church that day, I felt more peaceful than I had in all the twenty-two years since that horrendous day when my two friends had been dragged from life. I had gained an understanding and now felt a determination to finally put this whole, wretched episode of my life to rest.

My world was filled with memories and guilt that had succeeded in wrecking my life everyday, memories which were presented to me in the minutest of detail wherever I was and whatever I was doing. Memories of the screaming of a dying man and the sight of battered bodies torn and shattered by the violence of one single act dominated my waking nightmares. It was a guilt that extended to everything I did, sapped me of energy, and, at times, my will to live.

Over the years I had become to feel guilty about everything I did. I felt guilt if I drank to mask my memories, and felt guilt if I didn't drink. I felt guilt if I was enjoying myself and guilt if I was in a bad mood. The guilt in me welled to an impossible level as I considered the impact my irrational behavior had on my immediate family. My two daughters were too young to understand why their father, the man who had come to flit in and out of their lives with disarming irregularity, had been incapable of telling them why he was so irrational.

Whatever the circumstances of the accident on the glacier that day were, I had been used to living with danger. I had been a mountaineer and I had been working in Antarctica, one of the most dangerous places on earth, but throughout the years I had allowed and was still allowing images recalled in one minutely short period of my life to dominate my every waking moment. I had pleaded with myself that surely I was stronger than this. Surely I should have been able to thrust these thoughts back into the recesses of my mind and out of reach. Or better still, shouldn't I have been able to brush off the accident? There was something I had not been facing in my memories; there was something else that had refused to let me go.

The guilt I felt had forced me to make one massive mistake. I had refused to talk to anyone about the accident. People knew about the accident. The expedition organizers had informed my family within days of it happening, and friends had read about the death of two explorers in the press. I had attended inquests in Port Stanley and London. But still I could not talk about that accident on the glacier in the way I now realise I had needed to. The memories remained locked behind a door, and I had refused right of entry to anyone until the year before the visit to the church.

I had let down my guard for the briefest of moments when traveling in my car with a client of my business. The client had been questioning me on my thoughts about the meeting we were heading for, when after a short period of silence she suddenly asked, "What's the problem?"

"There is no problem," I replied. "The process is straightforward . . ."

"No. Not with the meeting," she interrupted me. "With you."

I knew in an instant what she had been talking about but I could not answer. My silence had been the key that proved to her she had been right and she persisted, refusing to let me off the hook. She continued, "You have something on your mind that you won't talk about, and I think it's something that's causing you terrible grief."

For the very first time, someone had asked me directly about the accident, and I responded, allowing the flood of my anguished words to wash through the inside of the car with an intensity that shocked her into stunned silence. I remember her swivelling around in the passenger seat to face me as I let go with a tirade of anger directed at the lost years and the shock of the sight I witnessed in the depths of that crevasse. For the first time I had allowed someone, a stranger, to share the pain I had suffered, and with my deluge of words, I had felt the weight of despair being eased from my shoulders. The client told me weeks later that she had forgotten about the problems she had with her business that day as she worried about and lost sleep over the way she had forced me into a corner. But as I looked back, I had felt it was to be the beginning of the first stage of what was to be my road to recovery.

Now on a dismal November morning, as the church disappeared in the rain splattered rear window of my car, I thought I had broken through another impossible barrier. Perhaps, the first stage had ended. Now I was beginning to talk in even more minute detail about the story. I was beginning to unburden myself truthfully, and with the unburdening came not only the usual painful flashbacks, but more detailed memories long since thrust into the back of my mind. Now, because I had finally been able to talk, I was beginning to understand, after twenty-two years of torturing myself, that the accident had not been my fault.

In the early years, a good ten years after the accident, I was offered help from someone who knew the barest details of the accident in Antarctica. That person gave me the heartfelt advice of disposing of all my possessions from that time to rid myself of all the hurtful memories. So, in pain and turmoil, I built a great bonfire, and piling it with remains of Antarctic clothing, diaries, photographs, letters, and telexes, I tried to purge myself. The bonfire was an act of cruelty, an ill thought out attempt to dispense with my past in one, quick burning, gasoline-saturated pyre of memories and treasures. It was a deed that was to create more pain, as the burning of my diaries meant my process of remembering was to become a series of further violent struggles.

Some photographs and a hand crafted wooden box were the only physical proof that had escaped the pyre. But there was also a small black book I found some years later. Discarded in an old tin box, the book had spent most of its years covered in dust and dirt in a disused cellar. A small reminder of times long gone, it was a record that had once been my daily sledging notebook. On coming across the book late one afternoon, I sat amongst the piles of old garden tools on the dirty stone slabs of the cellar floor and I read the contents for the first time in more than twenty years. This book had been my constant companion during my days of traveling in Antarctica and was kept in my inner breast pocket whenever I ventured from the security of the base. On its pages, recorded in pencil, were sightings, landmarks, positions, weather reports, detailed notes, dates, and times that had formed the basis of my longer reports that would, in turn, become my diary. I was shocked that the notes in this little book proved to me that the years had not taken a massive toll on my returning memory. Tucked away amongst the pages of the little notebook, I also found a couple of sketches I had made during the last enforced lie-up we endured in the tent. Drawn in pencil on the lined paper of the little notebook was Nick, the only other survivor of that journey, with only his nose and eyes appearing over the top of the protective neck of his sleeping bag. And, some pages later, the confused mess of a storm-ravaged campsite.

I now consider the wasted time over those years since the accident as I struggled to come to terms with the futility of my life. My love of the mountains and of mountaineering had been taken from me, and although I did climb again in the years before I began to suffer the horrific memories, it seems evident now that it was without the passion of earlier years. Ironically, some of the most difficult Alpine climbs I ever completed were during those years. The impressive mountains allowed me my time on their faces and ridges, and did not punish me as I perhaps should have been. The massive exposure and the consequences of the mighty drops into the valleys below were overshadowed by my thoughts of what might have been. Bereft of happy thoughts, I regretted the disrespect I had shown towards those sacred places, and I did not bask in the characterless, empty triumphs.

Gradually during the weeks that followed my journey to the church, I was drawn to the eventual outcome of writing down my story as a final attempt at unburdening myself of the guilt that had racked my mind for so long. When I began to record my story, and as I sat in my long sought for seclusion, I released my grief into written words and the memories poured from me, often being followed by my tears onto the page. But, at long last, with the writing came an agonized understanding. An understanding that I had been unable to grasp throughout all the years of torture and grief I suffered as I tried to come to terms with the events of that one single day. I had lived, others had died. My oppressive guilt had found its origins in that one single fact. But my guilt had been constantly fuelled by my employers from that time, who have failed totally, in the years between that day and now, to officially inform me if any blame had ever been attributed. Their complete silence encouraged my guilt to grow to the point where all else was unimportant. By their total lack of support, without them instigating even one offer of help during the intervening years, they contributed massively to my continuing concerns over the accident on the glacier.

Joe, Ron, Nick, and I had been thrown together as members of an expedition. I do know that those traveling companions and great friends, Joe and Ron, died following their true dreams and living a life they had freely chosen in a most difficult and dangerous world.

My thoughts still constantly turn to the glacier with the name that not only described its geography, but with the name that also described my life in the aftermath of my last visit to its surface. I now know that my spirit died with my friends on that glacier and I will never be free of the memory.

This is my story of the journey that ended uncompleted and in such tragedy. I wanted to record the events in a way that allowed me to purge myself once and for all. If some place names or dates and times are still not absolutely correct, then I can only apologize; this is my story only as I remember it.

2
1986 Flashbacks

As I played a volley low at the net during a tennis game one glorious summer's day, my thoughts were suddenly flooded without warning with snow, ice, cold, and fear. Usually I could admire the grace and skill needed to play a volley, but this time as I tried to punch the ball back at my opponent, I thought only of the carnage I had lived through. I thought of the blood streaks on the iced walls of the crevasse.

It was a short, massive shock, with thoughts that entered my mind for no reason or cause that I could think of. I was playing tennis in the almost unbearable heat of a Greek summer and in total contrast to my experiences in Antarctica. It was five years after the accident.

The tennis game was finished soon afterwards and I forcibly pushed the whole episode further back into my mind to avoid using my lack of concentration as a limp excuse for my limp attempt at winning the game. But without knowing it then, the next and worst stage of my life had begun.

In the early days, in the first five years following the accident, the memories of Antarctica would come to me with the sound of a two-stroke engine, a song, a smell, a cloud, or a breeze on my face. At that time, they were not painful memories; they were just memories of times gone by, both good and sometimes not so good. These normal memories would come and go as the cause, the sound or the smell, was recognized and connected with an event that had taken place during my time in Antarctica. Then, with a crash like a huge building collapsing, the newest stage had started during a tennis game. Everything suddenly changed. The increased intensity and aggressiveness of the horror of that time, long past, fought to dominate my life.

The flashbacks began to take place with a more purposeful regularity five years after the accident and a few months before my hurried return from my work in Greece to the UK. With a sudden start something would come into my mind and I would wander, lost in thought, as I tried to understand why it was happening. It was almost as if I was forcing myself to delve further and further into the depths of my mind to try to re-connect with hidden truths. I felt I needed desperately to relive the events, as I felt at that time that I could barely remember much of what had happened, but with the pain of remembering came long forgotten things I wanted to try harder to forget. During the more painful flashbacks, I would seem to physically connect with an event, which took place at the time of the accident on the glacier. These flashbacks were more dangerous to me than a normal memory stirred in a normal way, and they would cause me to regress into deeper and more troubling times. Everything I connected with in its turn had a connection with something more sinister, and it became an increasingly more difficult fight for me every time this happened to try and tear my thoughts in another direction.

Lying on a beach in Mykonos in the heat of the Aegean summer, looking at the blue green sea, I thought only of a sea crowded with icebergs and surfacing whales. The cold was transported over thousands of miles to me as I blotted out the sounds of the people around me on the sun-drenched beach. I could see, unmistakably, the Weddell seals hauling themselves up through their breathing holes they had chewed into the sea ice, where they would in turn lie down on the cold ice to be warmed by the sun in the sub-zero temperatures.

Clearly I saw another seal being shot to provide meat for the dogs. The man holding the gun was to play a massive part in the escape to safety after the accident, and the whole sequence of events would be played out in my mind for the millionth time. I could remember how this man had thrust a welcomed cigarette into my mouth as his first act when the rescue party reached our almost inaccessible place too high on the mountainside on the fateful glacier. I could remember how he submerged himself into a sea of depression later that winter when, in the aftermath of the accident, seclusion and a lack of understanding caused his sudden withdrawal. And I could remember the debt of gratitude I owed him for his bravery during the days following the accident and for the dangerous miles he had traveled to help us.

The memories of Antarctica were flooding back with relentless, heart-rending precision. I no longer needed a stimulus to cause me to regress into another time and another place where I could only think of death and destruction. Behind the closed shutters of my bedroom I feared the quiet time before sleep enfolded me. I worried why my perspiration would soak my bedclothes at night, even as I slept in my air-conditioned apartment. But my state of turmoil was even more difficult to bear as every morning I struggled with the pain of deep seated and nauseous headaches and the feeling of unbearable fatigue.

Five years after the accident my feelings had changed in an instant, in the time it takes to strike a tennis ball. Any satisfaction I had experienced in Antarctica suddenly became a great burden to me. My boxes of carefully catalogued photographic slides became repellent, quickly gathering dust as they began to lay unused and unseen for almost twenty years. I no longer wanted to admit my involvement in having traveled and wintered in that remote continent, and I would become aggressively dismissive if the subject was raised. My back was turned on a major reference point in my life as I tried to detach myself forcibly from my past.

What had been left of my good memories totally disappeared, pushed to the innermost recesses of my mind, more or less without a trace. I could no longer find the motivation to draw on the remembrances of the camaraderie and the magical experiences. During the increasingly difficult times that were to follow, I no longer felt any affiliation with anyone or anything associated with Antarctica, as I totally avoided any contact with past friends of that time, and wished more than anything that I had never set foot in that place. I was staring into a massive void.

One part of me told me to overcome this weakness to go back to the mountains and climb, to get back into the snow and ice, to hear again the chink of metal against metal on my climbing harness and move forward, aggressively thrusting the miserable thoughts from my life. The other part of me was telling me to keep my head below the parapet and gain the advantage by not drawing attention to myself. I desperately needed to lose myself totally in normality.

My story of Antarctica was now firmly behind a closed door, and although I hated having the key, I allowed myself through the door while it remained firmly closed to everyone else.

For fifteen years, these aggressive flashbacks tortured me as I tried to break free from the confines of the environment in which I was now living in the UK and which I had begun to hate with a passion that was dredged from the depths of my soul. But the environment afforded me a type of security, and I had not had the courage to turn my back on the hell of which was a massive part of that security. That cost seemed far more acceptable than being totally alone with only what remained of my memories and the onerous guilt that were my constant companions. As I fought my memories I needed peace and quiet, not the constant discomfort that my life had become. I was living a hell on earth and I became dependent on alcohol and painkillers as I tried to hide and then to kill my innermost feelings of despair. My pleadings to leave and move on, to somewhere quieter and more suited to those conditions I vitally needed, were dismissed. So I struggled on, becoming a virtual hermit enclosed within the walls of my home. My ability to cope was eroded, along with my motivation to succeed in life. I was caught in a trap, restricted by immovable forces against escaping in one direction, and faced with the loneliness and despair I would have to face even if I were able to escape on my own.

But eventually I found the will to escape, and a few weeks later I faced my impending loneliness through my car windshield. Behind me the place I had called home for many years disappeared in my mirrors. In my panic and confusion, I had no thought of the sadness I may have been causing my wife and two daughters by leaving, as my thoughts were totally filled with the memory of that time so many years before.

As I drove north in my heated car that evening, all I was aware of around me was the wide, open space that was the snowfield in Antarctica. My flashback was in full flow and it was the morning after the blizzard twenty years before, and I was immersed in the positive response as the quiet of the new day had filled our worlds. The blizzard, having raged for six days, had ceased without the hint of even a murmur. A dim, cloud-blanketed light and a massive silence pervaded our presence, a stark and almost unbelievable contrast to the screaming winds that had threatened us so aggressively and for so long. Relief from the storm had been tangible; the dispelling of the uncertainty of the situation that had gripped us for so many days came with a renewed happiness. Immersed in freedom once again and in control in its tight, secure confines, we had the last day of our journey to complete. Now there were no stresses, no complications, only the sight of the barrenness of the landscape and the replenished camaraderie that the difficulties had diminished during our confinement.

The digging out of our tents and equipment from beneath the hard packed snow began in a flurry of optimism. Our bodies were tired and drawn after the days of inactivity and the denial of sufficient food, but the thought of movement towards the safety of our base created energy and rekindled our joy. We knew there was a supply dump quite close by and certainly within a mile of our current position. But searching for the dump, now only a cross on a map and long since buried under the years of blowing snows and all but forgotten, was not worthy of long consideration. The effort required for the search would be too great, and there was no certainty we could locate its exact position or that, when located, its store of paraffin and other supplies we desperately needed would still be there. We had to move; that was the best option open to us, as the desire for safety drove us on, and with depleted bodies we packed our equipment in the urgency to open the throttles of our vehicles for the homeward surge that should take only one day of travel. We would follow the only direction toward and then down the shambles of a tortured and angry glacier; a route passaged many times by me, by us, and by countless others.

The sudden change in the weather had stunned me. An almost eerie silence now pervaded the landscape. It was a silence that had shocked me in the early evening darkness of the previous day, and allowed everyone our first relaxing sleep for a week. The noise of the storm that had engulfed us had been, at times, so deafening during the last six days, that to even think had been an almost impossible task. We had been threatened by a massive wind with its ferocity of aggressive speed and pounding weight for all that time. A wind that had suddenly ceased in the blink of an eye. A wind that had stripped two of our party of almost all their hope of ever getting back to safety. A wind that had kept four of us huddled together in our close confinement, barely able to move. A wind that had trampled our orange-skinned tent relentlessly. The poor, beaten tent that had barely protected us and kept us safe, the only sign above ground level on this vast and barren snowfield.

The early morning weather had been warmer at minus five degrees centigrade and I had crawled out of the tent and stepped into a virtual heat wave. We were immersed in totally new conditions that were unusual for this time of year. After having experienced major drops in temperature since our retreat from the cape, we were used to the exceptional cold. Now as the dawn came upon us, the sun shone through the retreating cloud and immediately I could feel its burning presence as it found exposed flesh by bouncing its rays from the individual snow particles in its effort to gain even greater strength.

Emerging, subdued from our confinement, we had to present ourselves to our tormenter; with pride we had to show we had withstood the severity of the storm and we had to show we were still intact both in body and in spirit. We could beat this thing now. We would travel away from this scene with straight backs but respectful of the immense power we had experienced.

3

The Journey: First Days

The vast, almost featureless ice piedmont had drawn Joe Watson and me to its emptiness. We had carefully planned our journey to the cape, and it wasn't a big deal. Four days hard travel north by skidoo, the motorized transport that had largely replaced dogs, across the piedmont from our base at the southern end of the island on the Antarctic Peninsula. The days were going to be shortening quite soon to the point where the sun would disappear for the winter period of darkness, and this would be our last chance for some months of getting a good trip under our belts. We traveled for the passion of traveling. Both of us had been beginning to get claustrophobic again in the confines of the base leading up to the mid-winter madness and we needed the release that was solitude.

The summer personnel had long since left on their arduous and anti-climactic journey home, and with them had gone some of the closest relationships in which I had ever been involved. The memories of times enjoyed with some of those people would remain with me forever, but those friends had now disappeared over the sea in the ship heading north, back to their own civilization. Of those friends, some would be affected forever by the grandeur and sheer unadulterated beauty of this vast and hostile environment. Some of them to write up, and continue, their scientific work. Some to return to their studies or their careers, and yet others to take the next steps in their hoped for return to this continent that had meant so much to them.

Those of us who were to stay for the winter had clambered down the scrambling nets from the soon-to-be-departing ship into the launch that was to return us to shore. Inevitably, amongst the excitement, there was the regret of knowing that those departing and those staying may never cross paths again. But now the new relationships amongst those left behind were being formed and the reality of the fast approaching winter had filled us with anticipation, as the opportunities for travel appeared boundless. Thoughts of mountaineering and traveling with both motorized and dog sledges had excited me, and I was keen to experience as much as I could in the short time before the winter season took the sun away from us. Although we had jobs to do, time was never going to be a problem, and I poured over old sledging reports and maps to view the possibilities.

As Joe and I pounded north on our route to the cape, the scenery was outrageously beautiful. The huge whiteness was spread before us, pulling me towards it, enticing me to step upon it, and taunting me with its sexy "come and take me" look. Occasionally, the ever changing conditions would bring us a period of whiteout, those times when the sky and ground merge into one white mass and the horizon disappears, allowing only severely limited visibility, and very often stopping our progress altogether. Sometimes, a brief but strong wind would slow our progress as the snow swirled in its spindrift, equally limiting our visibility of the massive whiteness and the mountains beyond that stretched for miles in all directions.

In my traveling partner I had drawn one of the many aces from our small deck of cards. Joe Watson was the base cook, sometimes quiet sometimes not, but interested in all that was going on around him and immensely interesting as a person. He had stories to tell of his apprenticeship as a professional soccer player, and I had known he was a real soccer player from the very first time I saw him kick a ball, those months before in the Falkland Islands. Soccer's loss was our gain, and Joe conjured up outstanding meals even when we were exposed to the vastness of the ice plateau at minus thirty degrees centigrade.

Joe and I had been on a number of trips together earlier in the winter and he was eager to learn about traveling over this dangerous ground. We were bound by the common bond of knowing that we both wanted to experience the vast openness of this massive continent and bask in its freedom.

The passage over the first days to the cape had been without any significant problems. Only those associated with the usual interruptions found when traveling in those conditions in this place. We had made our passage up the first major glacier in reasonable visibility and very low temperatures, and there had been no wind as we made slow and careful progress around the ice blocks and open crevasses as we gained height over a three or four hour period.

This was a magnificent place. Mountains to our left were casting their huge and daunting presence over us. Steep slopes to their heights were guiding us up the correct path to the top of the glacier, and the open crevasses were getting larger and larger as we traveled around the external curve upwards against the glacier's downward progress. To regain some warmth, we stopped for a short tea break, and took some photographs of one of the most massive, bottomless fractures I had ever seen. The day turned into a mini expedition, as Joe had never seen the inside of such a large crevasse and he was desperate to gain the experience. He tied on to a rope attached to one skidoo and carefully lowered himself to rappel into the beautiful, and strange abyss for the first time. He was astounded at the depth and the deepening blue colors as the crevasse disappeared below him. I followed him on my own rope, watching his confidence increase as he glided downwards, taking as many photographs as he could during his progress. We hung from our ropes at a depth of about sixty feet, relaxed and looking around us, and Joe, savoring the experience, took it all in his stride.

When the climb out from the blue crevasse began, Joe jerkily moved upwards and he cursed the fact that he was

beginning to sweat in his heavy clothing. He used his jumars, the tools used to lock onto and enable the climb of
the thin climbing rope, with a degree of skill that did not
suggest the tools had remained attached to his harness more
or less unused until now. He quickly mastered the skill of
self-rescue in this controlled environment. As we hung
from our ropes just below the lip of the crevasse, waiting as
we cooled down a little before exposing our sweating bodies to the colder air on the surface, Joe commented on the
hardness of the ice that formed the walls of the crevasse.
The ice was as smooth as silk and as hard as iron, and so
cold that it felt hot to the touch. Thousands upon countless
thousands of years of accumulating ice and we were touching it, caressing it with our fingers and wondering what had
been happening in the world when this ice had originally
formed.

Our reverie was interrupted with the usual expletives of
young men in such a situation, as we returned to reality and
considered the climb we had to complete. Gloves were replaced, our hands now throbbing in the cold, and Joe prepared himself for the final pull up his rope and the final lunge
that would enable him to regain the surface.

The experience added to Joe's understanding of Antarctica. He was no longer just "base cook"; his experiences had
surpassed the confines of the base area. The secrets of glaciers were being unlocked and he knew intimately the feel
of the cold walls and the deep blue color of the ice. More
importantly, he now knew that he could climb out from the
depths of a crevasse without needing my help. His experience also made me feel slightly more comfortable, as the
experience of this one short excursion had made Joe more
aware of the dangers and therefore safer in himself and safer
for me as a traveling companion. I thought back to my tu-
tor at college who had recounted a story of his training in
Glen Nevis, Scotland in winter. His party had been told to
dig snow-holes for their night in the extreme temperatures
of the Scottish winter, and when they had finished and made

themselves comfortable, lounging out of the increasing darkness and increasing cold, the instructor had shouted at them to get outside once again. "In normal circumstances," he shouted, "you'll be cold and tired when you have to dig a snow-hole, so cave that one in, get tired, and dig another one!" As I filled two cups from the thermos flask I could not help thinking that Joe still had a long way to go.

As the steam poured from our cups, we finished the quick brew of tea needed to get the circulation back into our bodies before we mounted our skidoos and once again prepared to head upwards on the trail up the glacier. An incident free passage, and we gained the point where we would turn sharp northwards once again, but now to enter into our virgin territory. On previous trips together, we had turned westwards and then south towards a deserted base one day's travel away. On this occasion, we were geared up and prepared fully for a longer and more difficult journey across the extensive snow-field separating us from the cape. Due to the conditions and the shortening days, we calculated four or five days of travel at most would bring us to our objective. The journey would be followed by a few days of rest and recuperation on the snow-covered and windswept area of the beach, before the return journey that could take a day or two longer, due again to the shortening days.

Our first night's camp was in an open nothingness of snow and ice, a short distance from the top of the glacier. The deteriorating weather had prevented us from making navigational sightings and the snowfield stretched eerily into the murky distance as we took advantage of the enforced stop, lounging and reveling in the silence. Not even a bird's noise shattered that silence, and we were too high up on the plateau and too far away from the sea to entice any other creatures. It was deathly still and becoming increasingly colder.

Drawing slowly, ever nearer to our goal, we traveled in the imagined warmth that we created with our enjoyment in the bitter cold of the plateau. The mountain we aimed for was far off in the distance, fifty miles of hard travel further on.

Joe, feeling the urge to break the monotony, kept shouting at it loudly from behind me, raising his voice well above the noise of the combined skidoo engines.

We traveled almost directly north for three more trouble free days, with the mountain that marked our path only absent when the whiteout visited us. The presence of the massive peak did not ever seem to get any closer as we fixed our sight upon it and never deviated from the course I had plotted leading straight to it. Our camps were set up in the vastness of the ice piedmont, sometimes during the early part of the day, as the weather kept changing and halting our progress, and sometimes just before it became dark, after having moved the camp soon after putting up our tent as the weather cleared. It was a thrilling journey of snow with the whiteness spreading out, without the hint of the telltale depressions caused by a thin layer of snow that signified danger below, hiding the opening of a crevasse.

Eventually we began our detour around the base of the mountain, altering our route to pass well to the west in order to approach the cape from the safest direction. On the windswept, open landscape we pulled over the last small ridge and into position at the top of the snow-covered beach that marked our arrival. My prized camera was out of its bag in an instant and the timer set to record this moment for our posterity. With broad smiles we stood, side by side, staring into the camera knee-deep in snow, and with one skidoo tilting away from the increasing wind, as we tasted euphoria in that place.

With a blast that shattered the silence we had only just ourselves disturbed in our exuberance in arriving, the first major storm of our journey appeared from nowhere. I didn't even see it coming.

As our cameras captured the landscape, the wind had crept up on us, and within minutes we were firmly in the grip of a blizzard that threatened to blow us off the face of the earth. We were going to be removed from the planet without a trace, and it felt like a punishment blasted onto us for daring to put our footprints in this long unvisited place.

Quickly we grappled for the security of the tent, and Joe had his first experience of putting up our shelter in a blizzard, as we continued to smile at our success while the reality of Antarctica was swiftly thrust upon us. Concern gripped me as I realized that in terms of "big blows" (blizzards), this was the biggest I had ever experienced by far. But with the concern came the overwhelming need for control, and all the procedures and necessary jobs were undertaken and completed in panic free enjoyment. We worked together as though we had been working together for many years. Joe produced teamwork of a master class, as he raced around tying down equipment and tent guys without any instruction from me, moving the right boxes to the right place, and unloading the sleeping bags and inside equipment as he battled in the teeth of the storm.

Within fifteen minutes of the battering beginning, the tent was up, the sleeping bags installed, and the primus stove was roaring. And still Joe laughed; I could hear him clearly even above the massive noise of the blowing snow, laughing out loud inside the tent as if he relished the danger, swearing at the blizzard to try harder if it thought it could beat us. I could hear him shout and laugh above the noise of the wind from my exposed position as I gripped the furiously flapping tie-downs for the covers of the vehicles to stake them out. Within forty-five minutes the site was as secure as I could make it. I cast one last look around the scene outside into the decreasing visibility, and then I dived through the now opened tunnel entrance of the tent, straight into the luxuriating warmth of the interior as the wind trumped up in its fury and the force grew to new heights. Joe was on top of the situation, as already a hot drink was made and food had miraculously begun to cook.

But the roaring of the primus stove quickly became almost totally drowned by a lunatic wind that screamed at the top of its lungs and breathed its impressive breath at the thin skin of our tent. I readied all the survival equipment as the severity of the storm continued to increase, and I feared the tent could be blown away in an instant as we ate food that had never tasted better. Joe, as he frequently seemed to do, produced a bottle of wine from within the folds of his bag, and even though the cold had taken its toll and frozen the contents, we were able to salvage enough from amongst the broken glass to toast our success. With the remnants of the wine evenly divided into our soup-stained tin mugs, we lay back on our sheepskin mattresses and began to enjoy listening to the wind going through its avant-garde choir practice.

Our regular radio schedule with the base was difficult that evening. The noise of the storm did its utmost to drown out the radio operator's words from all those miles away. After the shouted conversation was completed, we settled down, fully dressed and prepared for immediate evacuation, and tried to get some well-earned sleep in the grip of that raging storm. Fourteen hours later, as though it had never happened, the storm had passed.

Morning brought with it the shock of the devastation outside. Vehicles and sledges were buried in hard packed snow and the prospect of a prolonged dig out hit us with the force of the previous night's aggression. The speed of the arrival of the storm had not allowed us to get the most sheltered site for our few days of holiday, so we decided to move half a mile or so inland to a more protected place. With its usual sobriety the digging out began. Joe began the task of clearing and packing the contents of the inside of the tent as I sweated with the effort of removing the massive amount of snow from our equipment. I uncovered the sledges first to salvage some needed gear before turning my attention to the skidoos, but it wasn't long before the first threat to our journey was discovered. One vehicle was damaged to an extent that surpassed the contents of our meager toolkit. The right caterpillar track was shredded beyond repair, and the only solution was the complete replacement of the damaged part. Although we had achieved the cape without suspecting any damage, it was obvious that the track must have been torn before we had stopped the day before. I pondered worryingly over the chance of a repair and considered trying to drive the vehicle out to safety, perhaps even getting it to the top of the glacier. We could either leave it there or sit it out while waiting for a pick up or a spare part and tools to be ferried out to us. The thought of trying to wrestle with the damaged vehicle for four or five days, although depressing, seemed worth the try.

But the skidoo engine turned over lifelessly in the cold as I pulled repeatedly on the draw cord. It disagreed with my initial heartfelt pleadings to make the noise of an engine and I had no alternative but to strip the fairing from the vehicle and delve into the workings. Particles of ice had formed in the air intake and the carburettor, and after thirty minutes, as my pleadings turned into violent threats against the metal monster, it did eventually splutter reluctantly into life. The engine noise improved as it heated in the freezing air, and the skidoo simply tried to go around in circles when I sat astride it and put it into forward gear. The rubberized caterpillar track on the right side was catching and squealing as it dragged against the chassis, and it became quickly apparent that travel over all those miles on this vehicle was an impossible task. I now knew we had to get ourselves closer to the base with only one skidoo, and within minutes the decision to abandon the damaged skidoo had been taken. Our holiday dreams shattered for the present, we dug out the remaining equipment. We loaded only a single sledge. Our single skidoo would have struggled to pull the weight of two fully laden sledges with the additional weight of an extra passenger over the distance and terrain we were traveling.

Sadly, we deserted the site after I had manoeuvred the damaged skidoo and the redundant sledge into position side by side. Some full jerry cans of fuel were left in place on the sledge along with some spare boxes of non-essential equipment, and after covering the depot with a tarpaulin we made the site as safe and secure as possible and partially covered the mound with snow. Within a few days the sad depot would appear as a small mound of snow, but it could easily be picked out on the largely flat and featureless area of the cape. It is always a difficult decision to leave a damaged vehicle in the field, but in the circumstances, it was the only option we had. Optimistically, I decided the situation fed my desire to travel, as it would enable me to return to this point to retrieve the skidoo within a few weeks or when the sun returned after its winter break.

The orange fairing of the remaining skidoo pointed home-wards, thrusting its nose towards safety as I stood astride the machine and looked around at our depleted train. Joe had taken up his position at the rear of the single, trailing sledge twenty feet behind and amongst our remaining equipment. Both of us had safety lines attached, and we were protected from the cold with insulated suits, hats, bear paw mitts, and goggles. Joe, in his more exposed position, wore a facemask to cover any visible skin. With that, our brief respite was over and we were now travelers once again.

Within hours, we were well into our retreat, heading south, powering through ridges of snow left as a reminder of the angry wind of the night before. I had already contacted a two-man field party who were several days travel south of our position and heading towards a rendezvous point we had decided upon, close to the top of the glacier. We would meet there and decide what the best strategy would be.

It would be three days before the two parties met, and not even two weeks before half of us would be dead.

4
Beginnings

The coal dust and claustrophobia of my grandfather's every working day and my father's early life was in stark contrast to the snow and seemingly limitless expanse I now experienced. It was four years since I had returned from Antarctica and I collected my thoughts and climbing gear, as I looked out across the boundless space spread before me from the top of the north face of the Matterhorn. As I sat in the sunshine and forced myself to eat my mind wandered and my thoughts traveled the years back to my childhood and those forces, which had shaped my life thus far.

I had been cycling to school as my examinations approached, and I worried that the coal mines or the shipyards were my only options if I were to leave school the following year. My world had been filled with competitive cycling, and only eighteen months before, at fourteen years old, I had raced under the hour for twenty-five miles on a misty, still morning in Yorkshire. A month earlier I had achieved second place in the British National Championships. I dreamed of my heroes of the professional cycling world, but knew in my heart that I could never hope to emulate them, as my mind turned again to the depressing thought of a life underground. However, I was to scrape into the local grammar school for a sixth form of confusing career opportunities before again scraping, this time by the skin of my teeth, into college to become a teacher. It was as a Physical Education student in the early 1970s that I was thrust into the world and into outdoor pursuits.

The balletic activity of rock climbing turned my head, and soon I discovered the big mountains of the Lake District in winter. My makeshift clothing and old boots did little to keep out the wind and cold, but I was smitten. In this newly discovered reality of snow and ice, I touched the freshness of the scenery in the mountains. I witnessed the extraordinary view of the day's early light between the mountain peaks and found that no two sunrises are ever the same.

My introduction to a summer season of climbing in the Alps gave me my first successes on Mont Blanc, Zinalrothorn, and Barre des Ecrins. With a sun ravaged face, crusted and burned lips, but with my spirits soaring as high as the mountains, I had barely enough in funds to exist upon. The routes to the high refuge huts that acted as the gateway to the higher climbs did not prepare me for the shock of the sheer size and grandeur of the glaciers, peaks, and ridges in this new and beautifully hostile environment. Mountains poured onto me from every direction, towering upwards until they reached unbelievable heights. Mammoth mountain faces were interrupted with snow filled gullies, and enormous glaciers, adorned with house sized blocks of ice, cascades of stone fall, and jagged fissures filled my world.

My first ever bivouac above the Alpine snow line is a memory that will stay with me forever. As the sun disappeared from view it threw a blanket of red light over the mountains that coloured the vast snows and even the darker ridges and faces. Then came a perfect moon casting its own light over the area, as if granting us permission to identify the peaks we would see more intimately the following day. During the hours of darkness, as the drum roll of distant avalanches cascaded into my thoughts, I tried to get much needed sleep for my exhausted body.

My head-torch beam split through the dim light of the first hours of the very early day. Our small party of climbers trudged upwards through the frozen scene while shattering the fragile solitude of the outstanding beauty and peace with our sounds of exertion and occasional words. And in the distances, thousands of feet below us, myriad beams of bobbing torchlight betrayed further threats to the peace. The energetic first morning light cut through the cold with its new warmth, and I sat beneath my first Alpine summit. In the perfect setting for a breakfast of hot coffee and old bread, we waited as the sun broke into its full splendour before we emerged onto the summit. By the time the heat from the sun had turned the frozen snowfields into a warmer and more dangerous place to be, our ascent for that day had finished and we had raced back down our route to our bivouac site. Relaxing in the warmth of the early afternoon, lying beneath the presence of this great mountain, I luxuriated in the experience of the true camaraderie of this magnificent mountaineering experience with friends of only twenty-four hours.

Five years later, having experienced bleak but usually entertaining nights in snow holes and sublime days on long routes in the mountains of Britain and Europe, an advert for working with an expedition in Antarctica jumped from the page of a climbing magazine. Before having time to really think, I had applied, was interviewed, accepted, resigned from my job, and was amongst an array of people in one of the colleges of a University undergoing my induction in preparation for boarding the ship and heading south a month later. Incredibly laid back men, brimming with confidence, milled around chatting and introducing themselves in the filled-to-capacity lecture theatre. I felt like an intruder amongst these heroes and explorers of Antarctica, and during that first evening I realized the true enormity of what I was doing when I was introduced to a pipe smoking Sir Vivian Fuchs.

I had seen, while at primary school, the film of Fuchs's Trans Antarctic Expedition. Vivid recollections came to me of orange vehicles with caterpillar-tracked pontoons set at all angles struggling over crevasses. Of men, rugged and weather-beaten, hurrying to obey the orders of their illustrious and groundbreaking leader, Fuchs. Now dressed in thick corduroy trousers and woolly pullover, the great man drew on his pipe as he looked me up and down and we shook hands. In my awe-struck state I fumbled through my answers to his very straightforward questions.

I was asked if I had ever sailed. "Yes," I responded with trepidation, "I've sailed quite a lot during the last few years, mainly to France and Ireland." Amongst a group who joined us, another chap revealed when asked the same question that he had just returned from his third Atlantic crossing. "What routes have you done in the Alps?" someone asked. An opportunity to impress, but before I could respond someone else replied that he had just completed the Eiger North Face. "How long did you take?" someone else asked. "Seven days," came his reply. Someone else said, "I did it in four." "Yeah," he replied with a smile on his face, "but I did it in winter."

And so it continued. I was introduced to men who had spent ten winters in Antarctica, people who had just returned from the Arctic, climbers just returned from four months in the Himalayas and the Andes, and two chaps who had just returned from an assignment in Antarctica digging out a crashed airplane. Later, standing on my own with a drink in hand, I found there was still more to come when I was approached by a very big, and fit, looking man.

Our eyes locked as he strode purposefully over to me. Handshakes were carried out as the man stated simply, "What's happening." I blurted out my name as I winced from the pain of the handshake and he made some comment on the proceedings. This mountain of a man was not particularly interested in talking about himself and he was relaxed, imposing in his collarless shirt and close-cropped hair, which accentuated his impressively muscular physique. Looking back across the years, I realize there was never any doubt that the world would hear about this man. His journey from student, to tree surgeon, to budding Antarctic traveler, to one of the greatest polar travelers of our time, came in just a few short years.

5
Journey: Retreat

Rapidly decreasing temperatures saturated our retreat from the cape as our single vehicle traveling party ploughed its way south towards our rendezvous with the advancing two-man party on the plateau close to the top of the glacier. Spirits between Joe Watson and me remained high as we split the biting wind with our vehicle and sledge, and the beautiful scenery seemed to pour onto us as we moved towards home. The holiday may have been cut short, but the situation had produced its own compensations with its reward of a breathtaking panorama in the clear and freezing weather. Mountains and plateau joined together in majestic beauty with low cloud occasionally blocking the lower slopes of the mountains from view, only allowing the craggy and snow covered heights to be revealed. But for the most part, visibility seemed to stretch forever above a low swirling wind that only showed itself in the snow that it occasionally blew around us up to waist height.

Climbing the mountains, which bordered us to our left, had probably never even been seriously considered, and most of the peaks were not even named. No one may have set foot on their enticing summits. But I was certain that in more accessible places in the world their ridges, faces, gullies, and summits would have been eroded by the countless passage of leather boots shod with crampons. I could imagine the metallic noises emanating from their slopes as parties of windproof clad climbers battled upwards, aggressively attacking the snow and ice with their axes, making their belays and clicking their karabiners onto ropes. The mountains were stunning and I could clearly see the most exciting routes, with the remoteness and to a lesser extent the cold the only barriers to their virginity.

Although the engine of the vehicle did its best to drown out all other noise, I could still here the grinding of the rubber tracks attacking the hard packed snow in the clear, cold air. Whenever I traveled by skidoo I sat half astride my vehicle with my right knee on the seat and my left foot resting on the left footplate. I can still feel the comfort that the position afforded me, and easily I could turn my head to the rear and check that the sledge was intact, with the passenger aboard and still being pulled along in the wake of the skidoo. With the flexing and the rolling of the sledge Joe was finding it hard work staying aboard, and with the irregular ridges of hard packed sastrugi, the iced up remains of wind blown snow, the task was even more difficult. Nevertheless, Joe was still in place, hands gripping the ropes holding down our equipment, his body rolling back and forth as he absorbed the shock of the bouncing sledge. Above all, as he pointed his face towards me, he demonstrated by removing his insulated covering momentarily that his smile was still visible. We were firmly involved in a journey to safety now, a situation that, although still enjoyable, required determination to get our vehicle depleted party back to base. As we ploughed on I stayed focused on the well-being of Joe and myself, constantly searching the ground ahead for dangers and constantly scanning the horizon for weather changes. We kept a steady pace, no faster than a normal jogging pace on good ground, slow enough to be safe and fast enough to eat up the miles to our rendezvous.

Throughout my time in Antarctica, the best method to use to save ourselves in the event of falling into a crevasse while aboard one of these vehicles was a subject we constantly discussed. A rope between safety harness, which we wore over our outer clothing, and the skidoo was considered the best way, but there was little evidence to prove it would work. It was simply the prescribed means. When traveling behind a team of sledge dogs there was not the same feeling of insecurity. Part, or all, of a team of nine dogs stretched out in front of the sledge could disappear down a crevasse first, and being connected together could be hauled back to the surface, provided the sledge did not follow them into the hole. My questioning of a seasoned Antarctic traveler back in the UK, who had been an experienced dog handler in the late 1950s, on the subject of personal safety when traveling on a skidoo, had elicited a wry smile and a shake of his head.

"I often hauled my dogs out of crevasses. They (the dogs) would break through the snow of the crevasse bridge before the sledge went across," he had said. "On a skidoo I can't see that you would have much chance."

He had used his dogs as crevasse probes; I would be using the skidoo I was sitting on and roped to for the same purpose. Although travel with a dog team was restricted to a much lesser distance than during a full day aboard a skidoo, there was, in my opinion, a serious defense to continue with the use of these superb animals, from a safety perspective. On a fully loaded and fuelled skidoo with the additional weight of the driver and a combined weight of nearly five hundred pounds, there would be no warning. The skidoo would probably be the first thing to loosen the bridge of any hidden crevasse, and once the bridge was fractured gravity would take over. For the driver on board a skidoo, traveling could be viewed at best as hit and miss, and at worst as extremely dangerous and life threatening.

Crevasses are simply fractures caused by the irregular flow of ice over the surface of the land, which can be hundreds of feet below. Some crevasses can be massive and some of them are big enough to swallow a house without the building even scraping the sides as it free-falls into the seemingly bottomless holes. Such crevasses with massive openings were not as dangerous as those small enough to be hidden by a bridge, a thin covering of snow covering the opening on the surface. The odds were stacked heavily against safe travel. Man's need to travel more quickly made the faults in the system more obvious, and I thought back to my "official" training on the subject of travel in Antarctica. In the heat of the tropics traveling south aboard ship, I was handed my personal travel file from a pile being distributed to those of us who would specialize in travel. The file was brown manila, and it contained a number of type written sheets and badly drawn diagrams offering opinions on safety systems. Basking in the stifling heat, I studied the various safety systems for glacier travel aboard a skidoo, and Antarctica seemed a million miles away.

My personal ability at self-rescue, used for getting out of crevasses, could not be faulted. I had spent countless hours on the glaciers of the Alps practicing and practicing in all weather conditions. However, personal rescue used when traveling on foot in the mountains with a rope attached to another climber was a totally different subject from that situation now facing me. In Antarctica, I would be attached to a heavy skidoo and an even heavier sledge. The equipment could not stop me from falling if the skidoo fell down the hole with me, and I certainly couldn't stop the equipment if it fell into the hole without me—there were faults.

So in the stifling heat of the tropics I continued studying my brown manila file with its type written sheets and badly drawn diagrams of rescue techniques for Antarctic travel, and came to the same conclusion as many others had. It was by no means perfect, but it was the best available. Besides, I was young and I was used to mountainous and dangerous travel—what was fear? My desire and determination to travel across the vast barrenness of the snow deserts of Antarctica and to contribute to polar exploration dispelled all of my concerns. I approached the problems of self-rescue, which could be encountered on this hazardous terrain under the different and difficult circumstances that heavy vehicles could create, with a totally positive attitude. But I still asked searching questions of the more seasoned Antarctic travelers and heard only about the excitement and the constant need to be vigilant of the dangers. There were exciting stories of bulldozers breaking through the sea ice and sinking without trace, leaving the driver to swim for his life. Of Sno-Cats, the big four pontooned vehicles especially designed for such difficult travel, breaking though crevasse bridges, and of skidoos escaping by a hair's breadth. But always, everyone repeated stories and theories with a positive attitude and without fear.

My constant thought was, "Don't sneak up on a crevasse and frighten it into submission with the weight of the skidoo," as I concentrated on our route and kept a constant watch, while Joe and I ploughed through the little snow ridges and over the hard packed snow of the plateau. Steering clear of anything I considered a potential hazard, I dismounted time and again to probe the ground if there was even the slightest discoloration in the snow that could suggest a hole below. But the plateau was sound, and the terrain provided a safe surface on which to continue our journey.

Nearing the end of the second day on our retreat from the cape and during a regular break, I was shocked at the sight of my companion's brilliantly white-blotched face. Joe was not aware of any problem or discomfort, but the need to stop to get the tent up and get warmth into his body became the priority. Frostbite was in its early stages and although it was not a serious problem yet, it would get worse if the condition was not treated. We decided we would lie up for the whole of the next day. Laughing off the wound almost as a right of passage, Joe was simply getting used to newly discovered occupational hazards.

The mass of white spaced plateau offered us lots of splendid places to spend the night. But no matter where we decided to set up our tent, we were still many days' travel from the nearest camp site facilities. No wash house, no toilet complex, and not a shop to buy post cards and pay the campsite fees. So the tent that night, as every other night, was put up where we stepped from the vehicle, and again the interior was soon warm and noisy with the sound of the roaring primus stove.

The weather on that day had been truly magnificent and the traveling had been superb. It had been a great day amongst great days. The skies had been bluer than blue. The ice plateau was hard enough to allow positive travel with the outline of the mountain to our rear still throwing its imposing presence in our direction. As I stood outside the erected tent, with a cup of hot chocolate in hand, I looked around me from the featureless snowfield to the line of mountains protecting us from the east. Occasionally, wispy clouds formed, obscuring parts of the mountains from view, then they dissipated as quickly then formed again and another part of another mountain was lost.

If God could create just this one single place in six days, he had deserved his day of rest.

6
Meetings

The man mountain and I talked for an hour or so that first evening of our induction. He was unlike the others and I was shocked to hear that, like me, this was his induction. His relaxed demeanor had made me think he was an experienced Antarctic veteran. But all he stated was that the film *Scott of the Antarctic* had transfixed him, and although he had a thriving tree surgery business, which he managed to fit in around his incredible social life, he dreamed of going south.

Out of the blue, he said, "Look, will you cover for me for a while? I've got something to do." He read my thoughts. "No," he said, "not just tonight, but for a few days. I've got something to do in London." I responded, and he was gone.

I actually thought he was gone for good. I was approached a few times over the next twenty-four hours or so and asked if I knew where the missing person was, but I just shrugged; he was around somewhere. A day or so later, the man mountain reappeared, squeezing into the chair next to me in the crowded theater, and he joined in the proceedings like he'd never been away.

This was who he was. He showed a total confidence in himself and a disregard for anything that might waste his time. I began to understand that he possessed a determination to cram in everything that could possibly be crammed in to his life, and I learned new lessons about sheer single-mindedness.

On board the ship heading south, every afternoon during our weeks in the Atlantic, this future Antarctic hero would tie a red bandanna around his head, slip his headphones on, curl his top lip, open his diary, and write as the B52s blasted in his ears. Peace was spread like a blanket. Then, after an hour or two, all hell would break out once again, as he stopped the incessant pounding of music and pulled us along with his vitality and exuberance for laughter and life.

Very early in the trip, he had emptied his enormously heavy canvas kitbag on the floor of our cabin. Two huge dumbbells rolled onto the floor and his training regime began. Hours would be spent sweating on deck, in the cabin, in the hold, anywhere he could get the space to put his body through its torturous paces, and all the while the B52s continued to blast away at his eardrums. He put his body through excruciating pain as his muscles pumped to their maximum and he continued in his own world with his attention to physical perfection. At first some of the less informed members of our shipboard group had scoffed, but they soon realized that this man showed no concern, and was totally disinterested in their opinions, as the B52s continued to blast away.

I laughed as I sat with him in some South American bar as the girls lined up to feel his massive, rippling biceps, and he joined in the laughter with a roll-up cigarette hanging from his mouth.

Our group of friends sat in amazement as this friend defeated the massive "crew arm wrestling champion" while aboard the ship in the South Atlantic. With the contest in full swing, arms locked, muscles straining, and eyes bulging, he psychologically destroyed his opponent by rolling up a cigarette with his free hand and placing it between his lips before lighting it.

The madness of carefree youth continued aboard the tender boat ferrying us back to our ship lying in the Falkland Sound after we had spent an evening in Port Stanley. The night was very dark and the sea, for the interior of the Falkland Sound, was pretty rough, black, and with white caps rolling over creating a severe choppiness on the surface. We were standing together seeking a place in the line for disembarkation when he turned to me and said, "Look I want a swim, push me in the sea." I remembered, when swimming amongst the massive rollers of the Atlantic on the beaches of Ipanema and Copacobana, how he had gone out far farther than myself to catch the great Atlantic rollers of late winter hammering onto the beach. I knew he was a great swimmer, but this was madness in the extreme. So I pushed him over the side. Down he went, fully dressed, and with a splash he disappeared under the tender to re-surface on the other side at the foot of the ladder to the ship, and of course jumping the line for disembarkation at the same time. Calmly climbing, he passed one of the officers, who simply looked him up and down while shaking his head before he disappeared into the confines of the ship to retrieve his now wet Walkman and B52s tape.

In a departure lounge at Heathrow Airport some years later, a chance remark by a colleague, as a group of us were finishing our drinks before boarding our flight, sent me scurrying in search of a newspaper. "Well those guys got to the pole then," was all my colleague had said. I knew in an instant who it was. The newspapers didn't report any news of the journey to the South Pole, but I heard a snippet from someone else in the airport that a ship had been crushed by the sea ice and had sunk in Antarctica. There was now only confusion in the mixed messages I received, and the four hour flight was one of the most uncomfortable I had ever experienced. It was not until I could raise the world service on my radio that evening back in my apartment that the story began to unfold.

This man, this giant extrovert of a man, had dragged himself on foot eight hundred miles overland to the South Pole with two equally courageous heroes. They had faced grueling hardship that I never had, and never would fully comprehend. It was a magnificent feat.

Above all, in our close-knit group of friends who had traveled down through the Atlantic together, this hero of Antarctica became our figurehead of determination and great humor, and he became a very true friend to many of us.

Years later again, as I fought my own desperate fight with the bottle, I stood transfixed watching a television screen as this man talked about his personal life. He looked pained and sad, but I detected that huge mischievous glint in his eye. I could feel that his joy for life and living was still there, along with his monumental desire to seek total freedom at least once every day, and I could hear the raucous tones of the B52s being screamed out in my head.

7

Journey: The First Lie Up

Light came a little later that morning, and the sun tried to bring some warmth to our campsite in the massive exposed snowfield. I awoke to stillness and the complete, all pervading silence that I could almost hear. As soon as I regained my senses, I raised myself and glanced across at my partner. His face was the only part of his body showing outside the thick down sleeping bag, and although colored by the reflected glow of the orange tent, there was little evidence of the previous day's frostbite.

Joe didn't even mention his condition when he awoke; he just opened his eyes, pulled his arms out of the confines of the sleeping bag, scrabbled around in the bag close to where he lay, and produced a packet of bacon from some secret compartment.

Contemplating our enforced lie up, we sipped tea and watched the ice melt from the surface of our sleeping bags and drop from the walls of the tent. We both knew the stories of explorers traveling often for months on badly frostbitten feet and with frost damaged hands and faces. They had been true explorers. Our situation was different and the frostbite on Joe's face had to be given time to heal properly. The continued warmth of the tent, with the stove roaring throughout the day, would provide the perfect environment. Minus twenty-six degrees centigrade of intense cold was enough to keep me inside the thin skin of the tent, but a quick look outside confirmed that the weather and visibility were superb. Windless with the sun, lower on the horizon, blazing down upon the area bouncing its rays off the snow. Surrounding peaks were exposed in all their reigning glory and the sky only held very small wisps of cloud.

Eventually, amongst thoughts of deserting the now tropical heat of the tent, I began the long process of getting into my outdoor clothing. Thermal underwear provided a base layer, on top of which went moleskin trousers and heavy woolen shirt, followed by the insulated skidoo suit. Often, because of the very hard physical effort needed when working around camp, we would wear a thin ventile anorak instead of the heavy suit. The ventile anoraks were designed primarily for use when going through the hard physical effort of dog driving; it enabled ease of movement and reduced the problem of overheating. Footwear for traveling were large orange cloth boots with very thick soles and masses of insulation; difficult to walk in but absolutely vital when traveling in an exposed position on a vehicle such as a skidoo. Large bear paw mitts that extended up the arm nearly to the elbow, woolen hat, and goggles or dark sunglasses completed our apparel for these extreme temperatures.

Those of us who had brought with us specialist mountaineering clothing, discovered the only pieces useful to us when traveling by skidoo in these extreme temperatures were the one piece fiber pile fleece suits and jackets. Our down-filled duvet jackets were useful for lounging, and the newly manufactured plastic mountaineering boots some of us were experimenting with were kept purely for climbing.

The pyramid style tents we used dated back to the early days of polar exploration. They performed superbly in the direct force of the wind, and although there were stories of tents being ripped apart and exposing the occupants, I had not experienced anything quite that dramatic. The outer skin of the tent was an orange, high quality windproof material, and the inner was an off-white satiny cloth, which helped reduce condensation. Space inside between the two people in their sleeping positions was taken up with a line of wooden boxes that contained those pieces of equipment necessary for use inside. Cooking stoves, utensils, food, radio, medical emergency equipment, and spare clothing were kept inside, while fuel and traveling hardware were left outside, usually placed around and on the large skirts of the tent along with piled snow to increase the tent's stability. For prolonged stays in one place the tents were perfect, but much effort was needed to move the big lumbering masses of cloth and rope and the large aluminum poles.

For safety purposes we also carried with us a small two-man tent, which, although capable of squeezing two people in side-by-side, was uncomfortable as extended living quarters. In addition, everyone carried a two-man bivouac bag for use in extreme emergencies. This bag was a heavy nylon envelope, which could be staked out on the snow, and two men could climb inside with sleeping bags and mats. By closing the top after entering, body heat would help to keep the occupants warm and alive in an emergency.

While Joe continued his period of recovery inside, I set about tidying up the site outside. Boxes from the sledge were lying in the snow where they had been left, and the slight wind that had visited us during the night had proved sufficient to cover the sledge and skidoo with a thick layering of snow. As was the custom, the vehicle had been refueled the previous evening but it needed checking over to establish that our one remaining skidoo would be fit enough to get us out of trouble and back to safety. I whipped back the pull cord and the whole, beautiful area for miles around was subjected to the silence-shattering roar as the engine warmed up in its mistaken belief that it would be carrying us further on our journey that day.

Tasks take that much longer in the temperatures of Antarctica. The big gloves we wore, the layers of clothing, and movement in the snow all contributed to the extended process of sorting gear, but within a couple of hours the site was tidy and ordered. Without the intrusion of any wind or any other noise to interrupt our conversation, Joe, inside, and I, outside, continued to chat throughout the morning as though we were standing shoulder to shoulder at a bar.

Every food box we carried contained a variety of foods exactly measured to keep two men going for a particular number of days. Amongst the contents was a number of the dreaded meat bar, a dehydrated foil-covered block which, when unwrapped, looked not unlike a very large stock cube. No one ever admitted to really liking this preparation and it was eaten usually as a last resort after all the other de-hydrated foods were exhausted. However, my companion, a cook of some quality, always contrived to turn this usual mess into a meal worth eating and I did not fear the thought of having to eat meat bar when Joe had prepared the meal. As always from somewhere amongst the confines of his bag and as if by magic he produced some luxury. Tinned fruit, another bottle of wine, bacon, and even tins of beer seemed to find their way into Joe's possession, and he added a new dimension for me in eating well when traveling.

We lounged for the remainder of the day and as night drew around us, I checked the frostbitten face of my com-panion again. He had achieved an almost full recovery and we agreed we could travel the next day. So we settled down to talk and read and write our diaries and to while the rest of the evening away, in preparation for the rest of the journey. As I think back to those days in the tent with Joe, I regret the loss of my diaries more than at any other time. They were personal journals that no one else ever read and thoughts and personal feelings were recorded in this security. In the confines of a tent, particularly when danger threatened, they were a means of confiding innermost feelings and inevitably they mentioned tent colleagues and those we were traveling with. At times they were harsh and filled with criticism, a direct result of the circumstances and the exposure to other people's habits and at times, peculiarities, as the small liv-ing space was shared. Joe and I had had our disagreements when on base, but up until a week or so after the enforced lie up for his frostbite, we had never had problems when we were together in the tent. Largely we got on well, and there was always the opportunity to go outside if the mood took us, to escape for a short time from the close confines of our living space. He depended upon me to see him through our journeys just as I depended upon him for support and his ability to work the inside of our tent. I could never have written in my diary of a reluctance on his part to travel or enjoy our situation as I knew that was never the case. Our relationship was built on a strong foundation of mutual trust and we genuinely seemed to enjoy each other's company when we were traveling. When I think back to those times I realize that in the circumstances we had built a remarkable relationship, it was not as if we had been life long friends who one evening in the pub had decided to travel to Antarctica. We had been thrown together as employees of the expedition, and although Joe was vastly experienced in his job as base cook, he had had no experience of traveling in such a harsh and unforgiving environment. He adapted well to the hardships he encountered, and what he lacked in technical expertise and experience he made up for with his joy of life, his outrageous personality, and his seriousness to learning about traveling in Antarctica.

8

Traveling South

Our journey from England to Antarctica, by ship through the Atlantic, had been long and largely spent settling into some kind of routine. We would work for half a day on the ship every day, scrubbing the decks, chipping old and sea-water scarred paint from the superstructure, repainting, cleaning, and even being true sea farers as we regularly took our turn steering the vessel. The other half of every day would be spent attending to our own, often outlandish, needs and activities: lazing in the tropics, laughing in the storms, and reading and writing the diaries that were to become such a huge part of our lives. Enthusiasm was high and all the assembled first time travellers to this region were keen to spot the first iceberg as soon as the Antarctic convergence approached, and to get ashore into the world in which we were going to spend the next part of our lives.

Outlandish behavior became a massive part of our existence, and like a bunch of kids on a school outing we were constantly making as much noise as possible. We no doubt annoyed those around us who were not actually included in our little group with our childish, badly thought out antics, but those not directly involved in our little group were probably acting just as badly. Years later I cannot explain the motivation for some of those antics and I am lost trying to find either rhyme or reason. How can I attach logic to being served our evening meal amongst towering icebergs by a scientist who, dressed immaculately, spent an hour pretending to be some deranged television personality? He fussed over us, treating some like lords and others like riff-raff, appealing to us for quiet so he could tend to the needs of his high-class guests with a squirming bow from his waist as he served and fawned over them.

How can I explain that we watched the film *Alien* at every opportunity we had? Word perfect in the script, we would will the ill-fated space crew through the impending disaster speaking their words before they were able to utter them.

And how can I explain the crazy parties we held? Sometimes with only two guests, formally dressed, or worse, dressed up in full Antarctic traveling gear in the heat of the tropical night. One such party had eight quests all dressed as "Rockers" with slick backed hair, sideburns, and tight trousers. The only casualty of this stage of our ridiculous behavior was a washbasin shattered by a sherry bottle for reasons that are best left in the past. Wilfred collapsed in giggles as he caused the damage, and our uncontrolled laughter suddenly swung into panic as we realized that within hours the autocratic first officer would be making his inspection of the cabins. Laughing, probably drunk, Wilfred spent the remainder of the night gluing the pieces of the basin back together and truly believed he could divert the prying eyes of inspection. Needless to say it did not go unnoticed and simply confirmed we had been totally and childishly out of control, yet again.

Amidst the madness cultivated on board by those of us who were heading south to over winter and to travel, the vastness of the seemingly limitless ocean created a profound beauty I had never before experienced. The scenery from the ship rails was becoming increasingly magnificent the further south we traveled. Cape pigeons in their droves entertained us for endless hours in the South Atlantic with their outrageous feats of aerobatic displays, while the solemn wandering albatrosses that followed us for days at a time became our constant shadows. These gigantic birds shadowed every turn the ship made. Minimum movement of their vast wings created graceful turns and they rode the thermals like the experts they were, while occasionally they would drop to the surface of the ocean to pick up the discarded food thrown overboard. Now all these years later I realize I haven't seen an albatross since those days, and I miss the regal displays they played out before their astonished audience.

With a surge the surface of the water broke, the dorsal fin and long back appeared, and I saw my first whale. In a curved glide it looked massive as it swam quietly by the side of the ship, but this juvenile could not prepare me for the sight of my first big whale, which was soon to show itself. We had been aware of the presence of the creature for some hours and had seen merely a hint of the sleek back a few times as the whale exposed enough of its body to taunt us while it took air and blew through its blowhole. More and more of us were spending our time leaning over the starboard rail, cameras in hand, waiting for the opportunity for a good photograph. Nothing could have prepared me for the sight of this massive leviathan as it decided to breach in the now cooling air of the deep southern Atlantic Ocean. The giant whale rose straight out of the water, and with its large left eye fixed us with its gaze. As I stared I could clearly see the scars and lines of innumerable battles around its huge exposed head, but the intensity of the hypnotic gaze the whale returned captivated me. We were interlopers in its sea, and as with a massive surge it appeared, a smile of disbelief broke across my face. The huge animal had crashed through the surface from the depths to gasps from its awestruck watchers. The whale seemed to hang in the air with a massive effort before it returned to the brilliant blue green depths of the south Atlantic with another huge splash and surge of foam. Having no doubt satisfied its curiosity, the whale took its leave to continue freely on its journey.

Perhaps the whale was not truly free as it swam, pursued by maniac governments who demanded the whale's meat to satisfy the needs of its electorate, under the guise of research. Suddenly more of us began to wear our "Save the Whale" t-shirts with huge pride.

Ironically, the first major landfall in the sub-Antarctic continent we made was at Grytviken in South Georgia, amongst the intolerable mess of a deserted whaling station. As I walked around the old buildings infested with rats, delivered here over the centuries as unwelcome visitors aboard the whaling ships, and as I saw the abundance of flat-headed harpoon heads in the detritus of this crumbling and deserted epitaph to inhumanity, I was hit by the arrogance and greed of man. For generations we had raped the seas of its most aristocratic creatures, the legacy of which was this place. This devastation of rust and heavy iron machinery, half submerged whale catcher boats, and the whitened remains of massive whale bones littering the shore remained as testimony to the massacre.

Wildlife in abundance surrounded the deserted whaling station, including birds of every description, and even herds of reindeer, imported at the beginning of the century to satisfy the whaler's urge for a more varied diet lived in the mountains. On the shores lay the massive elephant seals, while their newly born young wallowed in the mud and tussock grass, and King Penguins wandered around the quayside.

The high mountains covering the island provided our first decent outing in the snow and ice since boarding ship. And although we had been told quite explicitly not to climb or pursue any remotely dangerous activity, we set out to satisfy our need to feel the mountain air and get our ice axes and crampons cold once again. Some of us spent hours climbing up through the snow and ice, and I saw for the first time the great glaciers of South Georgia that were so close to the sea. We gazed up at the faces of the mountains that made this place a mountaineer's paradise and looked longingly at the summits, many of which remained unclimbed. We slid for hundreds of feet down snowfields, a technique which saved our feet from the steep, grueling trudge of descent before we climbed again to regain the height we had lost so comfortably and so quickly, and onto yet another unnamed mountain. It was a time of brilliant and unrestricted opportunity to taste the air in the highest places of the sub-Antarctic Islands. But that day was also not without incident.

Twenty-two years later I was standing at the bar of the pub in Nant Peris in North Wales when I came face to face with one of the constant companions on the journey by ship to Antarctica. He shouted my name across the bar, and even though he had spent the intervening years deep in the bowels of Africa, it was as though we had seen each other the day before. The first thing he talked of was the outing on South Georgia, and he reminded me of a story I had long since forgotten but which came back to me as if a switch had been flicked on. We had glissaded, a controlled slide using ice axes to brake our progress, a long way down a glacier leading to the sea and the whole party, some seven or eight of us, all bumped into each other as we stopped. One of the party unbeknown to anyone, had taken his ice axe loop off his wrist, and he suddenly began sliding seemingly out of control, rapidly picking up speed. As I was the closest to him, I set off sliding down the ice slope trying to increase my speed to draw level to his flailing body. When I did get close I shouted, "Use your axe." Looking across at me as he slid on his backside, he comically raised his gloveless hands and shouted, "I haven't got it," as his slide, which by this time was almost totally out of control, headed rapidly for the ice cliff at the bottom of the glacier. Within two hundred feet of our impromptu cresta run, we were close enough together to reach across the last few feet separating us. Willingly wrapping his arms around my legs as I rolled onto my stomach and dug my ice axe into the ice, leaning over it with all my weight gradually slowing our slide until we stopped yards from the long drop onto the beach where the Elephant seals were aggressively going through their mating rituals. We both laughed uproariously before my companion, in his droll Irish accent, said, "That was nearly one of the more serious cases of coitus interuptus for the seals," as we peered down on the huge back of a three-ton male elephant seal directly below us on the beach. Then he put the final full stop to the incident, as he looked up the glacier the hundreds of feet to our party. "Oh fuck, have we got to climb all the way back up there?"

On our way back to the ship we avoided contact with anyone until the gear we had needed on our clandestine outing could be taken out of concealment beneath our jackets and returned to our cabins. The act of this little foray had been so much the sweeter for the stealth, but never once did I consider at the time that those who had told us not to climb had still half expected us to do so.

The South Georgia scientists had held a long unbeaten record in soccer matches against visiting ships, but they succumbed to our aggressive and determined method of play. Although the snow was a four-inch blanket over the entire field, soccer shorts and shirts were worn.

The game was taken remarkably seriously as we kicked it out in the lee of the Norwegian church. The game did not produce a high standard of play and the losers did not take the defeat gracefully, which showed in the many sullen faces on the base that evening while we the victors reveled in our unbeaten record for the season of played-one-won-one. All those who played for the ship's soccer team were called to the radio operator's shack on the ship as soon as we restarted our journey towards the Falkland Islands. The ship's radio operator was soccer mad and his world had suddenly taken on a seriously optimistic view as he realized for the first time in many, many years that the ships team would be able to give The Falkland Islands team a good game. Maybe even win.

Preparations were made and the radio operator became even more optimistic when another and larger expedition ship was sighted in the bay as we entered the Falkland sound. Another ship, bigger crew and personnel, more soccer players.

We took to the field in the bright purple shirts and white shorts of the ship's colors on a windy Saturday afternoon. The whole population of the Falkland Islands seemed to be crowded around the touchline, snugly covered in weather-proof coats or behind Landrover windscreens. Between all of us, on our team, we possessed one pair of soccer boots and nine pairs of assorted training shoes. Our goalkeeper wore his big mountaineering boots, and as a tribute to the cold wind, he also wore combat trousers and thermal underwear. But as we sat in the center circle before the game began, drinking whisky from a bottle of Famous Grouse, we were confident in our soon to be revealed secret weapons.

Our opponents took the field in a line of light blue, wearing proudly the emblem of the Falkland Islands emblazoned across their chests. They not only looked down at us because of our position seated in the center circle, but because they all looked so fit and healthy and ready, fully prepared following their season of Islands league soccer.

We kicked off in the strong wind, and as in all good soccer stories, the first time the opposition touched the ball was when their goalkeeper retrieved it from the back of their goal net. They realized quickly that players with very reasonable pedigrees filled our team in the more vital positions. Our mountaineering booted goalkeeper had tried out as a professional with his local soccer team. Our back four consisted of two ex-professional apprentices and a former Southern League center back who spent the entire game trying to run through his opponent, in pure 1970s style. There were smatterings of county and club players throughout the team, and our prize out of position left-winger, was a former professional league goalkeeper who could run like the wind that seemed to be ever-present in the Falkland Islands.

Months of inactivity aboard the ships unfortunately had taken its toll and the Falkland Islands won by the odd goal after a spirited game. Motivation was nonetheless high and the ships radio operator talked excitedly about the "next time." "Who knows," he said, "with a bit more organization and some training, less alcohol. . . ." Our response to this plan was muted agreement spoken in the euphoria of the moment, but we all knew we were soon to board ship for the final stage of our journey south. Even before the bruises and raging muscle aches from the exertions of the game had the time to dull, we had spotted our first iceberg, as we steamed further south. Massive telephoto lenses were hurriedly attached to cameras and focused on the white mound so far off into the distance. Experienced Antarctic travelers who had joined the ship in the Falklands Islands simply lounged and raised their eyebrows at our innocence and waited for the ship to get closer to the real south before considering even venturing on deck to take a look.

Penguins and seals were suddenly in abundance as camera shutters snapped noisily and repeatedly. We were beginning to experience one of nature's last, great, and largely untouched wonders.

9

Journey: Pushing On

I had experienced cold, but never anything like this, as I struggled into my outdoor gear to leave the tent on the third day of our retreat journey. The penetrating cold was making preparation for travel impossibly slow. With the onset of the winter weather, the dramatic drop in temperature was casting a different pallor on everything on these great snowfields of Antarctica. This was the Antarctica I had been expecting and we were now experiencing the cold of the ice cap. It was a dry, deep cold, which exaggerated that which would ordinarily seem to be minor problems, allowing them, at times, to take on monumental proportions.

Responding to the increased cold, getting used to its intensity, takes time. But although we rose early on the third day of our retreat and wandered through our normally relaxed breakfast, I was shocked to find that by the time I emerged into the bright light of the colder day it was almost an hour later than planned. I moved laboriously through my early morning tasks, trying hard to pretend to myself that I was not affected in any way by the severe drop in temperature. But the cold was more obviously having an effect now, and longer time spent completing the essential jobs simply shortened the time available for us to travel during the remainder of the day. By 11:00 a.m. that morning, the time we usually aimed to leave by, I still was not ready to move, so we decided to have another hot drink, have an early lunch, and sit in the relative warmth of the partially emptied tent in order that I could regain my strength.

Our food intake, like all travelers or mountaineers in the cold of these conditions, had to include an increased amount of calories. Energy is burned rapidly simply providing warmth when traveling aboard skidoos in the full force of such weather. The clothing we used was not hugely better than had been used by the dog drivers of years before who moved more when motivating their dogs than we did when traveling aboard skidoos. Some months before I had read with interest an out-of-date newspaper that I had been given during a supply restock. An "expertly" written article proclaimed that people stationed in Antarctica ate more food and put on vast amounts of weight as a compensation for a lack of sex. I felt that the writer, in this attempt at sensationalism, was far from the mark. On leaving the UK the previous year I weighed in at a healthy 163 pounds, and as I now topped the scales at 203 pounds, I could confirm that not all of my weight gain could be put down to my desire for a night of passion.

It was after midday when we started our journey that day, and with the cold wind on my face I felt relieved to get moving once again. The powerful view of the mountains retained their presence firmly on our left and the massive snowfield spread out in impressive command in front. Within an hour we stopped to take on hot soup and hot tea. Although the sun was now hammering us relentlessly with its reflected light from the snow, the stop was only short as the cold soon bit through our clothing. Miles in the distance we could see the same snow scene and we ploughed on, shouting at each other above the engine noise. The wind picked up at times, and swirling and blowing only a few feet above the surface, it caused the disturbed surface snow to totally obscure everything below waist height. Frequent stops to wait for clear ground and to check Joe's condition slowed our progress to a crawl and we eventually decided to call a halt after only three hours of hard travel. Our distance covered was meager and the sudden drop in temperature would have been much

better weather for traveling with dogs, as the constant movement would have helped to keep us warm. But at least we
had made some sort of progress in our journey south, and
it was unfortunate we had not had the weather conditions of
the day before during our enforced lie up.

It had been our first bad day, and Joe crawled first and
exhausted into the tent to sort the inside equipment while I
followed sometime later equally tired. We willed the tent to
warm up so we could discard our heavy insulated gear and
laze while we ate in the stove's heat before diving into the
sublime comfort of our sleeping bags. Because we were able
to carry the extra weight on the sledges, and because it was
winter, we had both taken two sleeping bags for this trip;
one of these Chinese duck down monsters stuffed inside the
other was an unbelievable luxury that mountaineers carrying
their equipment on their backs could not afford.

Once again, a blizzard tore at our tent that evening, whipping up our world into a frenzy of blowing snow and sending me the uncomfortable signal to move outside and check
tent and equipment one last time. I went through the tightly
squeezed process of getting into my gear before exiting the
tent to satisfy my conscience. Visibility was nil as the snow
swirled around my face, lashing my hair against areas of uncovered skin with the force of a whip. I crawled around the
base of the tent locating everything by touch, as I was ren-
dered almost blind in the severity of the gale. All the tent
guys, attached to skis and ice axes, were tight and holding
as the tent settled into its stable relationship with the snow-
covered plateau. Our equipment boxes were piled onto the
skirts of the tent and provided additional heavy weight to
stop our shelter from being blown away as the thick blown
snow landed and packed tightly against the boxes and created even more weight.

As the winds blasted against our only skidoo I could feel
it rocking with the force, but knew that no matter how hard
this wind, it would not remove the vehicle to too great a distance.

The blast of cold air that shot past me into the tent as I reentered sent Joe further into the folds of his sleeping bag, and I slumped onto my bedding to strip off my outer clothing and get back into the sublime comfort of my sleeping bag. The wind continued to blow furiously throughout the night and each time I woke it sounded angrier. But it was a night without incident and I did not have to leave my sleeping bag again.

10

Landfall

Leaving the rolling and pitching of the heavy seas to stern, the ship entered a mass of islands and inlets in the non-uniformity of the northern Antarctic Peninsula. We sought shelter following days and days spent in the teeth of a raging unrelenting storm. After the first few hours of bad weather no one had seemed concerned about our upturned existence as the force of the waves battering the ship sent replaced articles straight back to their position on the floor. No matter how hard we jammed ourselves into our bunks, desperately seeking sleep, tempers became frayed as we were constantly thrown around on this south Atlantic roller coaster. This island inlet allowed us shelter from the wind, and passengers and crew had an opportunity at last to get some long needed rest and sleep. For a day we luxuriated, sheltered from the wind, in relatively calm waters as the ship gently rocked us back to good temper.

The inlet allowed us to re-charge our flagging batteries. We were astounded to learn that during one period of the storm, while forcing our way into the teeth of the wind at ten knots, twenty-four hours later we had been pushed backwards for quarter of a mile. The ship had rolled and pitched and the mountainous waves had constantly broken over the bows, and even over the top of the observation platform above the wheelhouse, the highest part of the superstructure on the ship.

It was an experience that compared little to that which our exploring forefathers must have experienced in the sail-powered boats which had come this way less than a hundred years earlier. The seamanship required to get our powerful ship through these waters was outstanding, but at least we did not have to climb the rigging in this weather to change sails. Moreover, the hull of our ship was an ice strengthened double skin of steel, and not simply wooden planks and a prayer for safety.

We had been sent as shore parties to a number of sub-Antarctic islands either to restock other shore parties or pick up gear, and every time the scenery became more and more rugged the further south we traveled. Getting ashore was an exciting process, as transferring to a small boat from a larger one in heavy seas was an act of good balance and fearless commitment. Wait for the sea to rise, lifting the smaller boat to within jumping distance while hanging over the guard rail, launch yourself, free falling, at the correct moment, into the small boat in its frenzy in the surging seas. It was important to hang on to the first thing you were able to touch after you landed: another human, the boat itself, equipment, anything that would stop your forward momentum taking you over the side of the now rapidly falling, smaller boat.

On Bird Island the fur seals were belligerent at the distur-bance of their mating activities, showing teeth and striking out ferociously as we danced amongst them. The chinstrap penguins reluctantly gave way, squawking their displeasure, in their smell of fish and excrement. With outstretched necks and open beaks, they tried to menace us in order to protect their seemingly inappropriate nesting sites. The Elephant seals rolled and stared open mouthed and aggressive, snort-ing their fetid breath, and daring an approach to be made within striking distance of their three-ton bodies. The Skuas dive-bombed and struck the heads of the human interlopers who inadvertently wandered, in their innocence, too close to the territories of these aggressive sea birds. But, out in the bays the occasional, demonic silhouetted figures of leopard seals weaved their mesmeric patterns in the waters, while they waited patiently, amongst the broken sea ice and kelp, for their lunch to unknowingly swim by.

As we constantly journeyed southwards, the weather became colder, the icebergs more frequent, and the remains of the winter sea ice broke against the force of the bows of the ship. The scenery increased in its beauty and its relentless march to inhospitableness, and I wondered how life could survive in such abundance in this environment of ferocious predators and unforgiving climate. But the wildlife increased in numbers calculated by the million. Huge colonies of Adelie penguins moved and constantly screamed and chattered. Weddel and Crabeater seals lay in small groups on the retreating sea-ice, and the air was filled with birds of numerous species, all intent on traveling somewhere in their haste.

Our first sight of the island, which housed the landing strip for our final flight to the base, hove into view. The sun was shining through the cold air as Morse, who was to be the Base Leader on our base for the winter, pointed out the landmarks to me. Morse was a well-experienced Antarctic traveler with a wealth of talent in both work and humor. In the distance, he pointed to the hut in which we would spend our time waiting for the airplanes, and in the foreground was the rusting tin hut with the flag of Argentina proudly painted on its side; that countries claim to this strip of land. Half a mile behind the blue and white horizontal stripes of the painted flag, there was a steep rise leading up to the mountaintop. This was the snow-covered airstrip. I stared incredulously as I considered that I had been looking at the wrong piece of ground. Morse laughed at my bewilderment.

"What? The slope there, that's the airstrip?" I blurted as I pointed.

"Yea," he replied. "Quite a tight approach to it; makes the landing tricky. But the takeoff is something else. The aircraft just drops off the end of the runway. . . ." And he tailed off his sentence and smiled as I tried to understand.

There was nothing I could say and I wandered off to grab my gear to go ashore.

The next twenty-four hours was spent unloading the ship and ferrying equipment up to the hut. Then we hauled the same equipment by skidoo and sledge up to the airstrip that was being prepared to take the first inbound flight. Activity was unceasing, as we had to unload all the stores before the ship departed on its journey northwards away from the constrictions of any sea ice that could quickly form this far south. We were surrounded by the massively impressive landscape, while the smells of the nearby Gentoo penguin colony filled our nostrils and our senses with fish.

Our entrance into the world of twenty-four hour daylight created its own problems. A few days before, we had had a discernible nighttime; now there was just bright sunlight, all day and all night. So, we set up a wooden sheeted cricket pitch in the knee-deep snow, and played cricket at midnight as we waited for the airplanes to arrive. We became used to eating sledging rations of dehydrated food and dehydrated onion flakes sprinkled on the rock hard sledging biscuits covered with almost rancid tinned butter of dubious age. And we became used to the seriously cramped conditions as twenty of us squeezed into the hut, built for half that number, and tried to sleep in the unrelenting bright sunlight.

Following days of constant equipment moving, cold, cricket, penguins, and hilarity, we heard the distant strains of the airplane approaching from the south. At first it appeared as a tiny blot in the clear blue skies, then a small red shape, followed by a burst of engine noise with its own special tone, and the first flight arrived. The bright red aircraft with skis hanging below the thin wings and twin engines approached the snowfield in an uphill direction. I glanced nervously at the impossibly short distance of the runway, and back to the rapidly approaching airplane. With a thud, skis hit snow. With an immediate reverse thrust of its two engines, the airplane stopped its sliding and slipping within yards of touching down as the snow, blown feverishly around by the power of the propellers, spread its spindrift over the airstrip. Once again, there was an immediate burst of activity as stores were loaded and the senior man in the party climbed on board next to the pilot, put the big headphones in place, and closed the door. Furious revving, a change of direction, a violent shudder as the airplane built its engines to maximum power, and like a shot from a gun it roared on its skis towards the drop to the sea at the end of the runway. It wasn't going to make it, I was sure, and the airplane simply dropped from sight as it ran out of ground. Silence. An interminable silence. Then, with a roar of engine noise the plane reappeared, pulling its weight frantically upwards and under full power. I could almost hear the pilot straining to pull back on his controls, while the passenger no doubt held his breath and his breakfast in his throat. And with a surge of power it won its fight with gravity and it was gone, flying into the distance to become a red speck once more. The airplane had only one passenger on board, and a further nineteen passengers waiting for evacuation. We could feel the bad news coming.

Only one of the two aircraft had made the flight from Punta Arenas in Chile. The other one was stranded on the airfield in that distant land waiting the arrival of spares after the long flight from the UK. Flying rules forbade a single aircraft from operating in this hostile place for safety reasons. If one aircraft were to go down there should always be another for rescue and back up; now there wasn't. The ramifications of this news were huge. No traveling could be undertaken, which meant that lots of scientists with tight time schedules would be kicking their heels on base. We were now stranded hundreds of miles from where we should have been and our ship transport was on its way heading north on another mission. Worse still, the sea-ice, once apparently retreating, was now reforming, and another base, even further north, could not be reached.

The return home for the personnel of that base after the winter would be delayed for months, and was particularly difficult for one of those who were still a hostage to the ice. The scientist had had to endure an unprepared for winter on the island, as he had only intended being there for the previous summer of work, but unusual sea ice the year before had prevented him from leaving.

So in our small hut, with our dehydrated food and little to do, we prepared to wait again until a solution was found. We returned to the cricket game, and we wandered around the penguin rookeries and took in the wonder of the place as we waited.

Relief eventually came with the sound of blaring sirens as another ship appeared steaming through the inlet a week or so later. We prepared to retreat from the island that had become our temporary home with all the stores we had so carefully ferried to the airstrip, and we worked frantically to get supplies and men aboard in record time. This ship, more equipped and bigger than the previous one, took us a few hundred miles south through increasing pack ice, and we were put ashore amidst further unloading of equipment and furious activity at a deserted base that was separated from our final destination by a mountain range of impressive beauty and height. No time was spared, as the ship had to escape the encroaching sea ice quickly, and we found ourselves in yet another new home again waiting for the aircraft to complete the final stage of our deliverance.

The reason this base was deserted, we were informed, was because the glacier on which the airstrip was situated had disintegrated to such an extent that crevasses littered the whole area. Largely, it was unserviceable and dangerous, but it was all there was. So we waited, again, and checked the runway for the best landing site available, and we waited some more.

Months of confinement in small cabins on ships and in huts and the frustrations of delay and lack of sleep were causing morale to fall. Some people felt their time was being wasted as they had only come down for the summer season, which was quickly disappearing. Increasing frustration at the inactivity was casting a cloak of dissatisfaction over the party. Grabbing the best places to sleep was paramount to those few, and I was pleased to get as far away as possible from the discontent. However, this time the wait was not more than a few days and spirits were raised once again as we heard the familiar sounds of the airplane approach. Early evacuation from our temporary home to the main base was not required as the other, damaged, aircraft remained in Chile and still no flights from base were to be allowed into the deserted sites for scientific work for quite some time to come. But in small groups we were transferred the remaining miles during the following week, and the deserted base was once again left with only the birds and seals to enjoy its seclusion.

11
Journey: The Rendezvous

The wind was still whipping around the tent as I began the long, uncomfortable process of digging out the equipment after the blizzard that had continued throughout the night. As a continued sign for the beginning of winter, the temperature now seemed to be on a determined fall and the first lungful of air had created the familiar early morning shock to my system, rasping at the back of my throat like a strong cigarette. Scattered around me were the all too familiar results of the night's blow, with equipment boxes buried under the snow and our skidoo slightly tilted in its resting place. Thin ice pointers, glistening in the early morning sunlight and lying almost parallel to the ground, were attached as a blanket of hoar frost to the exposed metal of the vehicle.

I found it difficult at times such as these, in the almost complete silence of the early morning, to believe where I was. As the noise from the tidying up operation inside the tent began, I put down my shovel and tied a safety rope to my harness. Struggling through the deep snow I began to walk away from the tent, up wind, to the full extent of the one hundred and twenty feet of rope. The view across the snowfield towards the glacier was an enticing, mile upon mile of continuing whiteness. The mountains, with their lower slopes as smooth as cream, lay on my left and the plateau all around me was, in places, whipped up by a wind that disturbed the surface but created clear unobstructed visibility only a few feet above. Although pristine white and not gray, the scene fashioned the eeriness of standing in a churchyard with the mist swirling around at knee height, only allowing the tops of the gravestones to peep through.

Nothing disturbed the scene. Nothing, that is, except my presence. There wasn't a blemish that I could see, not a sign, other than our traveling party, that man had ever set foot in this place. So little pollution, yet I knew it existed and I knew exactly where the nearest mess was. One day's travel to the southwest, standing alone on the ice field, there was the fuselage of a wrecked aircraft. Further south again but well to the east, there was the discarded aircraft and skidoo fuel dump. Hundreds of barrels, surplus supply of a recent expedition that had neither the inclination nor the funds to remove the mess, leaving it all as a reminder of man's desire to conquer and leave his mark. And of course there were the bases and the presence of man that inevitably meant the normal pollution of everyday life.

Unseen and new pollution lay on the ice left often by selfishness. Only a few months previously, I had seen an aircraft in the jungle camouflage and insignia of another country struggling to take off from our airstrip with its insides crammed with visiting scientists. When no progress could be made the pilot gave up his attempts to reach the sky and disgorged the occupants onto the airstrip. With the airplane now empty, apart from the pilot and co-pilot, the engines had little problem powering the takeoff into the clear blue cloudless sky, and at the height of a few hundred feet, gallons and gallons of surplus aviation fuel were jettisoned all over the pristine white snowfield. With the aircraft now so much lighter the pilot landed, picked up his human cargo, and turned northwards towards South America cheerily waving to us and without a further thought to the damage he had just caused. It had been an unbelievable act of selfishness that I found hard to comprehend. Snow and ice thousands of years old had just been contaminated by someone who was too selfish to find another way around the problem.

I remembered also an earlier conversation I had on the ship as I traveled down to Antarctica the year before.

Hanging over the guardrail only feet above the pancake sea ice and marveling at the silence even with the ever-present throbbing of the massive engines in the background, I was joined by a crewmember. In a relaxed peace we were watching the small blocks of ice silently moving aside to allow the forward progress of the great red hull and were shocked as a big black globule of ice and oil and dirt was caught in the bow wave of the ship. The dirty mass, only a few yards across, seemed to want to cling to the ship as the wash tried and tried to push it away. We stood silently watching the globule for a few seconds. Tattooed and ear ringed, the crewman had a reputation for being as hard as nails when he was sober and positively dangerous when he was drunk, and I had learned to give him a wide berth when I caught a glimpse of his rolling walk and his rolling eyes. But his appearance belied his sensitivity. "Leave this beautiful place to the birds and the animals!" That was all he said as he continued to take in the man-created mess of this small piece of appalling pollution. After a few more minutes of silently watching as the oil and dirt were eventually pushed away by the red hull, the crewman moved on, and I watched his orange-clad, boiler-suited back disappear through the forecastle hatch.

Abruptly, as I pulled to the full extent of the safety rope, it lifted from the snow and tried to turn me around as it restrained my moving body. Knee deep in snow I remained still, looking around at the mountains and plateau. Taking in the fresh cold air, I sought the farthest extent of the horizon. Although the snow now blew around and obscured anything below waist height, I could still see the full extent of the mountain range and the hazy extent of the plateau. Eventually, looking skyward, I tried to see if I could catch site of the vapor trail of a commercial jet that I knew that I would never see in this region. It pleased me to know we were too far south for that type of intrusion, at least.

My pensive moment was abruptly ended as the cold clawed at my still body, telling me I had to move now to get warm, and thoughts of digging out to prepare for the day's travel dominated once again. As I trudged back along my safety rope to the camp, I viewed the scene and grabbed the shovel as the noise of equipment being packed from inside the tent brought me starkly back to reality. I relished my short periods of time alone in the stillness of the early mornings. This time of the day remained for me the most beautiful of times as the darkness slipped away to reveal nature's nightly round of cleaning.

The travel throughout that morning was hard in the ever-increasing wind and wind chill and we took frequent stops to warm our bodies. It was more comfortable for me with the orange fairing of the skidoo deflecting some of the wind, but for Joe, hanging grimly from the rear of the bucking sledge, it was becoming progressively harder as the full force of the wind hit his exposed body. Even though Joe was wearing additional layers of clothing, both his own spares and a lot of mine, the cold penetrated even the best barrier and the stops became more and more frequent. During this time he never complained about the weather, the terrain, or the cold; he just got on with it and told me frequently how much he enjoyed the whole experience. He was determined to see more of Antarctica and felt that the back of the sledge allowed him an unrestricted view and that he did not have to concentrate on driving.

Only a few hours after we began that day, the weather deteriorated and the landscape became featureless with the mountains disappearing into cloud. In the poor contrast and increasing whiteout we stopped to put up the tent as we prepared hot drinks and cooked our lunch.

Inside of the tent we soon warmed up, and we were able to remove our outer layers of clothing and lounge contentedly in this little oasis of luxury as we ate. We talked of the morning and the conditions and how we felt we were going. We were not unduly concerned as the progress we needed to make was being achieved, and we had managed to keep close to our self-imposed schedule that was so needed if we were to meet up with the other field party before dark.

Checking the map against the merest hint of a landmark amongst the low cloud enabled me to establish our position so we could move onwards. The chilling cold was a sign for good safe travel over the hardened surface at a fair speed, and all in all things were looking pretty good, although the reduced visibility concerned me. We rested for a short while longer to take advantage of the last few minutes of luxurious warmth as the stove was closed down, and before the cold once again began to penetrate the interior of the tent. When we left the tent the cloud had lifted, creating almost perfect traveling conditions, and I took heart in the change that had come over the area with such speed.

During that afternoon the temperature must have risen a little as some of the small ridges of snow allowed us to pass through them rather than over them, and the spray of the loosened snow slapped against my goggles. Joe continued in good spirits, at times signalling me to stop so he could warm his aching limbs and fingers and toes for a few minutes. He jumped about in the snow, warming his extremities, occasionally ceasing his dance to drink from the flasks that were filled at lunchtime and were rapidly emptied of their hot tea.

Without sophisticated navigation equipment routes had to be found by using only a compass and map, and later that day I thought we were lost as I scanned the reappearing horizon through my binoculars. The other party was nowhere to be seen and in my concern I did several 360-degree sweeps of the area standing on the seat of my skidoo "Shit", I said to Joe, "what do we do now? Move on or wait it out?" We set up the radio but I could only listen to static and I could not raise the other traveling party. The break in the mountains that marked the opening for the top of the glacier had been clearly visible, and I was certain the other party could not have passed us. Eventually I came to the conclusion that they were hidden from view in some depression in the snow-field, caught in the decreasing contrast of the Antarctic late afternoon.

Around this time the other party was pondering exactly the same problem within a mile of our position. They were slightly closer to the glacier than we were and had decided to get their tent up and wait until I found them.

Joe and I had a quick discussion, and the engine was pulled back into life as we checked our watches to give it another thirty minutes of southwards travel. We moved towards the top of the glacier as the visibility improved, and then just as quickly deteriorated, causing us the added problem of whether to stop or go on for the full half hour of travel.

As the weather conditions taunted us with constant change, the time gradually played out without a glimpse of the other party. The prospect of spending another night with our limited traveling capability before making our rendezvous was looming, but if that were to happen, radio contact would be possible and we could use flares to get a fix.

With a shock, as I pulled over the top of a large snowdrift, I saw them. There in the depression was the welcoming site of the orange tent of Ron and Nick. My passenger shouted that he had seen them, and with thumbs up to Joe, I steered slightly to the right to head straight for the makeshift base. I breathed a sigh of relief in the intense cold, and as I approached the tent, Nick emerged with a cup of hot drink in one hand and a cigarette in the other, smiling his welcome. His smile turned to astonishment as he watched me drive straight past him, but instead of fighting with the iced up throttle lever, I pulled the string attached to the engine cut off and surged to a tilted stop, burying the front end of the machine in yet another snowdrift. Greetings were made and a mug of hot tea was thrust into my passenger's heavily mitted hands as he pulled back the insulation around his face to gratefully accept the warming liquid. Three of us stood in the open snowfield passing the steaming cup back and forth along with the cigarette, enjoying the hot liquid, smoke, and the welcome company.

Ron, the second member of the other party, was not there, and it was explained that when they realized they had set up their tent in a depression, he had gone off to have a quick look around after gaining a higher vantage point. He returned within half an hour or so in the deepening gloom and we set about comparing notes in our now erected tent. I had been disturbed to see, as Ron arrived at the campsite, the empty rear of the skidoo that was without any survival equipment on board. But I presumed he had simply followed his tracks out and then back.

The cold was sapping us all, and the tiredness of the journeys the two parties had undertaken was taking its toll. The situation was relaxed, but we understood the need for urgency to avoid being on the plateau for longer than we needed with too few skidoos and my party's depleted supplies. Both Ron and I were concerned that the weather was going through a big change. If we hit a major time-consuming problem, the equipment Joe and I had managed to carry from the cape could be less than would be needed and would mean that we could be under provisioned. Although we were only two or three days travel from the base, in the worsening weather we did not know how long the journey would take. My major concern was the amount of paraffin we had available for use to fuel the stove, as we had dumped some of the fuel to save weight at the cape. I was depending upon Ron's party having more than enough paraffin.

Unfortunately they were also running short, but a quick look at our other stores meant we would be okay for at least four or five weeks, as we could, if necessary, burn skidoo fuel, a mixture of gasoline and oil, in our stove.

The route back to the base was in reasonable condition as Ron and Nick had traveled up along some of its path to meet us at the rendezvous. We were optimistic we could be back in the confines of the base within a couple of days, and I relished the prospect of my return with spares and equipment to salvage the damaged skidoo from its depot at the cape before the darkness of winter was upon us.

We spent the early evening in one tent catching up on the limited news, and when the radio schedule was completed we settled back in the warmth as Ron tuned in his transistor. The Voice of America came to us over the airwaves, bringing with it information about the horrors of the Iran-Iraq war. I pondered how lucky I was and how happy and free life could be.

12

New Arrivals

The pillar box red aircraft almost scraped the mountains before sweeping low over the headland. Squeezed alongside equipment and personal survival gear, I peered out of the port side window as the aircraft zoomed over the heads of the over wintering base personnel.

The area around the huts on the ground below looked dirty as the retreating snow was leaving the usual flotsam and jetsam of habitation in its wake. Glaciers, sea ice, and mountains in their pristine coverings, provided a much better advert than man's attempt at colonization. Banking, the aircraft turned to the left and headed straight for a high rocky ridge bordering a wide-open snowfield. As soon as the skis touched the snow, the aircraft was shaken by the familiar sound of the propellers being pitched to reverse their thrust; the only braking system available on the snow field that acted as the summer airfield. We taxied, stopped, and immediately the distinctive engine noise was cut off and we poured out onto the safety of land and into the massive over-crowding silence.

Leaning against a big orange Sno-cat, a four caterpillar tracked, pontooned vehicle, the unsmiling driver was waiting to transport us the few remaining miles down the edge of the glacier known affectionately as "the Ramp." Men, mainly long haired, all scruffily dressed, who had had little outside contact for eight or nine months, looked and occasionally nodded, but said little in their reluctant welcomes, cautious of the intrusion yet frustrated by our late arrival. We did our best not to impose too much upon them until they had become used to the massive number of extra people in the invasion. We were gate-crashing their tight knit community of closed relationships and post-winter problems, and they seemed like cats spraying their private territory in this very private world.

Although assigned bunks, many of us spent the first few nights as far away as possible from the very close confines of the main building. We sought somewhere to sleep in peace and quiet amongst the mess of workshops, lofts, and outbuildings, and away from any potential confrontation we may have caused in our inexperience and exuberance of having finally arrived. The over winterers had received, by now, their first mail from home in more than eight months, and they needed peace and seclusion and we willingly moved away to allow them space. Any resistance to newcomers did not last long, however, and the over winterers' humor quickly embraced this enforced change as they had been embraced the year before when they had first arrived. Soon the base rumbled on in its preparation for the departure of the summer sledging parties for the advancement of scientific knowledge, which was, after all, the reason we were all there.

More problems, however, cast their ghostly cloak over the proceedings, as the absent aircraft remained grounded in Chile. Without aerial support the travel itinerary for scientific work was delayed its inevitable start. The base was now full to overflowing, and it must have been dreadful for those who had spent the winter in almost total seclusion and who were now eagerly looking forward to getting on the ship bound for home.

During the period of mid summer, Husky puppies were born, buildings painted, new enclosures erected, equipment checked for the traveling parties, and the base was bustling with activity for almost twenty-four hours every day. We had our jobs to do and we just did what we thought was needed in the purposeful yet relaxed atmosphere, taking time off when we wanted to and getting to know the immediate surroundings. When it became too noisy and too crowded, as more people continued to arrive and visit, we relished the prospect of appearing on the night watch rota to get some semblance of peace and quiet.

When, at last, the repaired aircraft appeared from its dormancy in South America, the base slipped smoothly into overdrive as the field parties were dispatched as quickly as possible. I had been detailed to travel with a scientist a few hundred miles south and I spent hours lying on equipment in the rear of the airplane as we approached our drop off point at around seventy-one degrees south. We were disembarked quickly on an expanse of land that had only been discovered in the 1930s, and our equipment was barely on the ice shelf before the pilot restarted his engines and the airplane powered into the sky once more for its return to base. We loaded the sledges and started up the engines of our skidoos, and I lead off to travel the last few miles to our intended site. As the scientist roared past me, his throttle wide open, over hugely dangerous ground, I slowed my progress.

I spent weeks in a tent with this particular scientist. During the long days, hundreds of miles from anywhere and in superb weather, the scientist repaired his constantly malfunctioning equipment or waited for primitive computers to provide data. He dropped probes into the seawater at the bottom of the holes we had drilled through the ice shelf and we waited for a response on the inked needles tracing water currents and their speed and salinity.

During the long periods of inactivity I read the collected works of Dickens, practiced my climbing on the vertical ice walls of crevasses, and gazed longingly at the mountains that bordered this remarkable place. In the heat of the midnight sun we could clearly hear the avalanches thundering down the mountains in the far distance. As the sun blasted down upon us day and night we often wandered around our campsite on the ice shelf wearing only tee shirts and jeans.

Relief from the tedium came when a field party stationed a few days' travel from our position journeyed to us to celebrate Burns Night, the tribute to the Scottish poet Robert Burns.

Meetings were hearty and jovial, and we all sat together on the ice shelf and ate a meal together with the mountains and open miles as our backdrop. Later, when the two scientists closed ranks and discussed scientific things, Wilfred the instigator of the celebration party produced a bottle of rum and we two moved away from the tents and lounged on our sheepskins on the ice. Well into the bright sunlight of the night we read the increasingly unintelligible Scottish verse and celebrated the life and death of the great poet, all the while passing the bottle back and forth. My visitor regretted that our drink was not Scotch whisky, but nonetheless it tasted good, and eventually the scientists joined us and the rum quickly disappeared in tribute to the bard.

Two weeks after the Burns Night celebration, disaster struck. The scientist with whom I worked drove his vehicle, with me on board, over an ice cliff on the way to his worksite a few hundred yards from the tent. Although not a huge ice cliff, only twenty feet or so high, the near whiteout conditions meant that the driver had no chance of stopping the vehicle in time. The landing was hard and the scientist escaped serious injury, sustaining only a badly bruised thigh. I sustained damage to my neck and we waited for an airlift out and back to base. Due to the continued whiteout, it was two or three days later before an aircraft could get to our isolated position and get us back to base. The results of the X-ray showed minimum damage to my neck and a cut up closed cell foam sleeping mat was used as an improvised collar. I was up and about within twenty-four hours after being briefed by the scientist's boss. The pain in my neck emphasized the need for good training and careful travel in this hugely dangerous place.

Further aircraft damage cut short the scientific traveling season, and all field personnel were put on standby for an early recall to base. The single remaining aircraft worked around the clock, bringing home all the widely dispersed field parties from their positions around the Antarctic Peninsula. It was a massive task carried out superbly by the support personnel, who only relaxed when the final airplane flight touched down with the final incoming party aboard. At the time few people discussed what the situation would have been like if both aircraft had been unserviceable, but the prospect of several scientific teams traveling hundreds of miles overland back to base without air support could have proved to be a highly interesting, yet highly dangerous, prospect.

The end of the scientific season evacuation loomed, and on base everyone was feverishly working to assist in the dismantling of the damaged aircraft and the restocking of the base for winter. As soon as the relief ship appeared, the end of the summer season was on us, and the ship laden with its departing summer staff and homebound winterers sailed out of the bay one bright cold day as the sea ice began to sneak up. It was little wonder that the northwards bound travelers were smiling as they hung from the guardrails. Firing flares and waving to us from the safety of the departing ship they alone knew of the mammoth mess they had left us. Our first week of isolation would be spent clearing up and trying to find the hidden tools and personal equipment we would need for the winter.

The vision remains intact, and I can clearly see the departing ship parting the waters and leaving its trail of disturbed foam in its wake, taking with the turbulent waters the last contact with civilization we would experience for many months. Soon the realization that nothing would enable physical relief from this place until only the weather and the passing of winter allowed broke through the bravado of youth. There was now no escape from the marooned existence, but we welcomed the prospect of complete isolation from the rest of the world, and I was filled with a greater enthusiasm for this place; a place so desolate that only a few months every year will allow intrusion. For a short period of my life I would be removed from newspapers, from television, from drawing a wage, from the clamor of traffic, and all the other aspects of life we so take for granted. It would be months and months before I would taste again fresh fruit and vegetables. Everything I would experience during that long period of winter was here amongst this small community at the bottom of the world. The people here with me were to be constant companions in this expanse of white for half a year or more. I felt totally free; free of the restraints of life in the real world, and I felt few complications could cast their shadow over us to spoil my winter in Antarctica.

13
Journey: Powder Snow

Side by side on the open snowfield, the two tents flapped noisily as they were buffeted by the aggressive, freezing cold wind. I was convinced we would be swept away over the mountains that ran in a line down the north east side of the snowfield. But the orange tent material seemed unaffected as it allowed the wind to punch relentlessly against it. Not having claimed the victory of destruction, the wind swept onwards in its seemingly erratic but never ending journey across the plateaus and mountains.

At first light we called between the tents. During the darkness it had been impossible, in the noisiness of the blizzard, to hold a conversation, and knowing that the other tent and crew were safe was a relief in the stillness of the blown-out storm. Ron and I broke through the entrance tunnels at the same time, having had to dig through piled snow for the remaining few feet before tunnelling upwards to emerge through our wind blown covering. The snow, a very dry, extremely cold white powder, was built up to a good four feet in the lee of the tents, and on the windward side it had almost totally buried us. We waded almost chest deep through the snow as we surveyed the scene and tried to make some sense of the deluge.

Little could be seen of our skidoos as a perfectly levelled layer of snow covered them up to the very top of the tarpaulin covering them. The boxes, left outside with the sledges, were totally obscured from view and it was going to be a monumental task to find everything before facing the daunting task of digging it all out. Three hours of laborious work were standing between us and the beginning of the day's travel.

The snow was difficult to walk through, and so cold and so light and feathery that there was almost a fear of inhaling its very fine granules. Traveling in these conditions was going to be difficult, but the skidoo train would have to make impressive progress to get off the plateau and down the glacier as soon as possible.

Three hours after beginning our digging out, all the equipment was loaded and we revved up the engines of all three skidoos. With perfect visibility and a temperature of below minus thirty degrees centigrade, I took a last look around and led the line of skidoos towards the direction of home. Bouncing over the snow and, at times, submerging the whole front end of the vehicle so that snow poured over the bright orange fairing and windscreen in a torrent across me at waist height, I made my first attempt at progress. It was worse than we had anticipated and within twenty yards my engine choked to a halt as it drew in the light powder through its air intake. A machine that had been designed to withstand extreme terrain and temperatures had stopped with the carburetor choked with snow before it had even had time to warm up properly.

Within minutes all three skidoos with stalled, air-starved engines were immobile within twenty yards of beginning the day's travel.

Nick, who was a mechanic, was confident that the engines could be restarted soon and within twenty minutes all three engines had been cleared and were carrying us forward. But, after a further twenty yards or so my engine cut out again and I frantically waved the others past in the hope that they would manage to get a little further. Within yards one machine stalled and the following machine crawled to a halt as it tried to cut a way through the almost impenetrable powder snow. A new strategy was called for and we extended our safety lines to enable us to walk to a point we could all reach to discuss what we could possibly do to make some headway.

As soon as my skidoo engine had been cleared again I unhitched my sledge and began the job of cutting the trail for a short distance. It was our feeling that with less weight to be pulled, and by standing at the extreme rear of the skidoo in order to provide the weight to lift up the front end of the machine, we would have a better chance. Cutting a trough, I ploughed through the snow, leaving the others to follow exactly in the tracks almost vertically banked on either side with displaced snow. It meant we would have the murderous task of ferrying the sledges back and forward along the track, but as we were only a few miles from the top of the glacier it seemed to be worth the effort.

There then followed grueling hours of broken skidoos, stripped carburetors, and submerged sledges. We manually hauled the fully loaded sledges through the waist deep powder snow to hitch them up to skidoos, and we moved masses of snow with shovels. In all, during the eight hours of daylight and the five hours of traveling that day, we managed to travel less than four hundred yards before we reluctantly decided to give in to the elements.

During that day we kept the excruciating cold at bay by eating as we worked, munching our way through chocolate bars to keep up our strength and warmth as we swore, cursed, and laughed at these horrendous overpowering conditions. The sun beat down on us undeterred through the clear freezing air, and the only exposed parts of our faces, below goggles and above our chins, were burned as the sun reflected off the millions of particles of snow built up to such a depth all around us.

We estimated it would take three or four days to get to the top of the glacier if Nick could fashion some sort of snorkel for the air intakes on the skidoo engines. We would be lucky to get out in ten days if we continued in this way.

As we approached the end of the traveling day, I looked around at our party and beyond into the very close distance at the imprints in the snow of the previous night's camp. Our party was too spread out now and in our tiredness we had allowed the distances between us to become too great. Nick on one skidoo was about fifty yards ahead of me, but Ron was fifty yards behind, and as he waved at me I knew that his skidoo engine had stopped once again. I struggled to turn my machine around manually in the high walled trench of powder snow before going back to the rear of the column. There was no point in trying to repair it now so late in the day and so close to stopping for the night, so we hitched the two vehicles together and pulled the tired machine up to Nick at the front of the line.

Camp was set up and we tied everything down. Joe cooked in one tent so we could sit together. Nick knelt in the snow in the biting cold, and with bare hands he stripped and cleaned carburetors yet again before we dived into the tent and finally ate a long craved-for meal.

Darkness was upon us in an instant as we sat and discussed our predicament, considering a lie up for the following day, as progress was almost impossible. If the wind was to come to our assistance, however, and wipe the powder snow away, we could probably get back on schedule. But as we hauled our beaten and fatigued bodies into our tents to get some well-earned rest, the weather remained freezing and perfectly clear with not the hint of the wind we longed for.

Joe and I lay on our mattresses that evening giggling in our tiredness while we continued to swear and curse at the conditions. We laughed some more when Joe foraged in his bag and like a magician held two frozen beers above his head in his outstretched arms. The sweet tasting American drink eased us into sleep, but I was awoken some hours later as I heard the wind almost imperceptibly begin to gather pace.

Around midnight we were hit by an almighty storm that raged for the next twenty-four hours. It buffeted the tent from every direction and screamed against and between the guy ropes and the equipment outside. As I forced myself outside during the night, fighting against the howling wind, I remained tied into a rope. My goggles froze over immediately and offered no protection as they were blown from my face by the violence of the wind, and I protected myself by crawling on my side with my mitted hands covering my eyes.

We spent the long hours of the next day in the restfulness of warm sleeping bags, listening to the howling of an angry wind separated from us only by a thin tent skin. The stove provided a warming glow to the tent for thirty minutes a few times during the day, and we were able to get out of our sleeping bags and relax in the added warmth as we ate our meals. Periods of long silence continued between Joe and I in the tent as we read, worked, and relaxed, often falling into sleep for short periods of the day. There always seemed to be something to do during the lie up times apart from reading and sleeping. Boxes had to be sorted, equipment checked, usually something needed repairing such as a jacket or a pair of wind proofs ripped or frayed by the hard work of traveling.

Cleaning teeth and washing were rare occurrences in these conditions as all water had to be made from melted snow, and we were loathe to waste the depleted stocks of fuel. Similarly, before the days of good battery operated shavers, shaving was avoided. Toilet requirements were carried out quickly when considering the ravages the bitter cold wind could cause to backsides bared in the storm. Human waste was not buried, because in this freezing place, anything buried would remain for millennia as a reminder of the person's visit. It was better to let the wind drive the snow clean once again and reduce man's impact on this magnificent place.

We were surrounded by snow that was so cold and powdery that it became a tiresome and massively fuel-consuming task to melt enough to make water for a drink. But it was all we had, and we needed to remain well hydrated, which proved a difficult task in this snow filled continent that remains, ironically, the driest in the world.

Joe and I both managed to finish our books that day, so were able to swap and continue reading more or less without a break. In view of what was to happen later, we both wished we had read a bit more slowly.

14

The Generator Mechanic

A robust figure with black tousled haired walked past me towards the generator shed. Hands thrust deeply into the pockets of his moleskin trousers, boots unlaced, he had a tendency to look down when he walked. His brow knitted, deep in thought, he strolled past in thought, blocking out the rest of the world.

I'd met Maurice briefly aboard the relief ship when he arrived, friendly and very confident; he showed a determination to take the fullest advantage of this experience.

He was the man responsible for the upkeep of the three massive Volvo generators that kept the electricity and heat going for the base. Guarding his generators with a deep protectiveness, he surrounded them in his palace of a generator shed with control created by hard work, tidiness, cleanliness, and complete order. He would spend much of his time each day cleaning the floor and his beloved machinery and tools when he was not up to the tops of his arms in pieces of Swedish machinery.

Slavish in his determination for perfection, he extended this passion to his skidoo or any other traveling machinery he used, always servicing the vehicle himself to make certain that it would perform to his exacting standards.

Throughout our many hours together, I learned more about the winter Maurice had endured the year before at another base. It was there that he had witnessed the death of a very close friend. Ironically, the man had been struck by the landing ski of a low flying aircraft. The pilot was completing the welcoming low pass over the base that marked the end of the first relief flight, at the end of the long drawn out winter. During one long night, Maurice and I talked together through the quiet hours at the dog spans, with only the dogs for company, and he told the story. We talked and we waited on the ice as the howling of husky's welcomed one of the last suns of the summer to dawn in a display of red intensity. The mesmerizing, streaking light of the new day played out as the mournful soundtrack of the dogs created a fitting homage to Maurice's sadness.

During our time together on the base Maurice and I climbed regularly together on the vertical ice cliffs facing one side of our base. We spent hours in the darkness of the winter days playing table tennis in one of the laboratories with an intensity that suggested both of us were afraid of losing. We constantly interrupted the carpenter when he was trying to work, although never so hard that he could not put away his tools for a time. We learned Spanish together and played in guitar duos and trios, again with Simon, the carpenter, with me dragging along in the rear trying to look competent, if not sounding so.

But above all, Maurice loved his Husky's. He drove his dog team with care and increasing skill and was the first and only person that winter to get his team running on a nine-dog fan trace. He had trained his dogs so well that instead of running them in pairs behind each other in one long line, they now ran almost free, attached not to other dogs but directly to the sledge. A feat that showed great nerve, confidence, and calmness, to which his beloved dogs responded wholeheartedly.

Throughout the winter the night watchman's job was a role most of us relished, as it was an opportunity to spend time alone. One of the jobs was to trudge the hundred yards or so up to the Stevenson screen, the white slatted box on stilts that housed the basic meteorological equipment, to take the necessary minimum and maximum temperature readings. During the winter months it was dark, and it was scary in the loneliness of the walk. The outlines of the black rock

crowded in menacingly as the Stevenson Screen was approached up a zigzag, snow-covered path. Occasionally the
dogs would howl for company from their spans on the other
side of the base, but usually the noisiness of the shattering
silence would dominate.

One particularly dark night, at the time of the midnight
weather reading, I was moving slowly to the Stevenson
Screen, and the hard snow was scrunching very loudly under
my double boots in the extreme cold of the moonless night.
A quick noise and I was knocked flat on my back before I
could react. Trying to get up quickly I was aware only of the
sound of footsteps racing away from my inert body. It was
too dark to see who had hit me with a perfect rugby tackle
that had sent me into the air and flying onto my left side.

Struggling to my feet I broke into a run to try to intercept
the retreating tackler, but intent only on pursuit, I did not
expect the second, harder tackle, which flattened me into a
snowdrift. This time the tackler laughed as he ran off and I
recognized the dulcet tones of the quiet generator mechanic
as he raced back to the sanctuary of the base. By the time I
burst into his room on base he was feigning sleep tucked up
in the folds of his sleeping bag.

Some days later our radio operator had complained that
communications with the outside world were causing a problem, and the night watchman shook me awake at 3:00 a.m.
"You'd better come outside!" was all he said, as the door
closed behind him on his way to wake the others. I hurriedly
dressed and raced outside to be met with the beginnings of a
display of the southern lights, an event unprecedented where
we were and totally unexpected by the meteorologist. The
green streaks of the aurora curtains were beginning to establish themselves in the sky to our west, and the pitch darkness
of the night only accentuated the amazing display that was
about to gather pace. All the assembled personnel raced to
grab their cameras and I managed to find my fastest roll of
Ektachrome film from the depths of one of my bags. We had
Ektachrome equipment and darkroom facilities on base and

I would be able to see the results a year earlier than with the normal slide film I had to send back to the UK by ship.

I ran a hundred yards or so away from the base lights, and I set up my tripod while beginning to explore the possibilities of time exposure to capture this phenomenon. A figure some yards away was similarly erecting his tripod and sorting cameras. "Maurice, what do you reckon I should use for a time exposure?" I shouted to our resident Ektachrome expert. He simply laughed and offered no help, obviously determined to capture the best image of this phenomenon.

The competition was on and I was not going to be able to gain any or even a little of his hard earned knowledge. I trolled though my mind drawing on my limited photographic experience. "Okay," I thought as I calculated f-stops and lenses. "Start at a twenty second exposure, then forty seconds, then a minute." Within twenty or thirty minutes the display had passed through a less colorful period, and we were left staring, once again, at a black sky.

We wandered back to the base and I began the process of pleading with Maurice to develop my slides in his most skilful way. He took my film from my outstretched pleading hand, tossing it up into the air and catching it a couple of times. He looked at me, grinned, and immediately walked into the darkroom and closed the door firmly behind him, as I tried to keep up my groveling monologue with him. I was now in the hands of this nutcase and totally at the mercy of his outrageous sense of humor. I began to pace the floor outside like an expectant father waiting for the birth of his first child at the hands of some deranged midwife.

Time dragged by amidst the noise of running water and the chinking of glass flasks from within the darkroom as I wandered the corridor outside drinking my round of endless cups of coffee. The results of Maurice's own photography were less than adequate and his attempts were thrown across the dark room. My Aurora Australis, in contrast, showed the display in all its riveting glory. Brilliant colors streaked across the sky with the outline of the rock and ice buttress silhouetted in the lower background. But best of all the millions of stars in the background added a depth and professionalism that I was usually totally incapable of. My offers to help Maurice with his photographic techniques more than made up for his rugby practice on my body in the cold of the weather reading walk.

I had introduced Maurice to climbing vertical ice, and when he decided to drop eight or nine feet from the ice cliff one day because he got stuck, it presented another opportunity for me. Looking up at his ice axes still planted in the ice out of reach above his head, he turned to me. "Steve, lend me your axes so I can go up and get mine."

"Maurice," I responded, smiling joyfully as I walked off laughing, "these are my axes; yours are up there." But he came back at me again when I was thrown from my sledge and the dogs ran the last few hundred yards back to their spans alone. The sound of Maurice repeating time and again, "Never lose your dogs!" continued for weeks.

Following the accident on the glacier, Maurice used his strength to get me through. We traveled together with our dog teams in the cold and crisp air of the Antarctic winter, covering hundreds of miles and reveling in the experience. The ground we traveled over was chosen carefully, and although dangerous perhaps by normal standards, it was safe to us then and assisted in my rehabilitation to life in the frozen continent.

Years and years later I received a package from Maurice that contained the black bound and hand written sheets of a diary he had come across, by pure chance, in a garbage skip outside a house clearance in Cornwall, England. Contained in the hundreds of hand written sheets, in ink and pencil, were the words of a now dead Antarctic traveler of the early 1950s. It confirmed that our explorations were not new, and it also told that our relationships formed in the harsh world of ice and desolation were also not new. This unnamed man's experience had been told in his eloquent and humorous way, adventuring into minute detail as he bared his soul to the privacy of his pages and rewardingly to us.

I had thought our high jinks as we had traveled down on our way through South America had been unique until I read the words contained in this diary recounting stories of ribaldry and young men's laughter. The writer recounted his story of an attempt to steal a policeman's hat by one of his friends in Argentina in 1951. The "criminal" had been apprehended and his punishment had been to sweep the streets of Buenos Aires for one day. Under close guard the convict swept, surrounded by his contemporaries who spent the length of the unfortunate sentence applauding his sweeping. They whistled, cheered, and derided his efforts as the day wore on, much to the amusement of the guarding policeman, making more work for the inmate as they dropped litter in front of his broom and in areas he had recently cleaned.

With great delight I discovered that the broom-wielding criminal of this story was the same man who had nearly thirty years later offered Maurice and I our jobs and the enviable chance of seeing Antarctica.

The writer had detailed the close relationships he formed and how he luxuriated in the warmth those friendships created in the bleakness of the Antarctic winter. Years later, I remember my true friendship with the generator mechanic.

15

Journey: Storm

Ron and I stood outside the tents taking in a totally different landscape to that which we had seen before the horrendous blizzard. The site was now all but totally clear of the pristine and vicious powder snow that had so marred our progress. If someone had been down wind of us at that time they would probably have had all that snow dumped on them to frustrate their own plans for traveling.

We began the long process of digging out equipment from under the pile of hard compressed snow to get the machines started and warmed up. The snow was packed tightly around the boxes and the tent, forced there by the relentless wind, and the cold early morning air had solidified the mass and it proved difficult to move. But we worked on; we needed to move from this site and the weather, although far from perfect, looked as though it may allow us a full day's travel. We consulted and decided that although it probably would not let us make it all the way back to the base, we could probably get down the glacier and onto the exit ramp near the bottom; the point where we turn ninety degrees south and climb steeply to gain the large snowfield leading to home.

We struggled on, moving the remnants of the storm with our attention being increasingly drawn to the ever-worsening weather breathing on our backs. Once the tent was down we again consulted about the darkening conditions and again decided that traveling was possible. So with piled sledges and everyone aboard the vehicles, we moved away from the site line astern, heading inexorably towards the glacier in an ever increasing wind and the beginnings of hard blowing snow.

Within a mile or two we could see that the weather down wind did not look too bad. Unfortunately, the direction we were traveling was succumbing fast to a big bank of very low black cloud, and after a quick break and sips of tea from the flask we decided to lie up for a few hours and review the situation.

The process was begun in reverse and we stripped a sledge of its large tent, erected it, and dived in for a meal and some rest.

Four of us crowded into the warm interior of one of the tents after securing the now flapping canvas, and we sat on boxes to wait it out. And we waited and we waited and we waited. And the storm grew stronger and stronger and stronger as the time slid by. With the increasing wind came the deteriorating weather as we continued to sit it out.

By mid afternoon it was clear that we were not going to be going any further that day. Ron and I thought that with the depleting opportunities for running the stove it would be better to share the large tent for sitting in and erect a small two man emergency tent for him and I to sleep in that night. The entrances to the tents were almost touching, so getting from one to the other would not pose a problem, nor would the occupants have to get fully equipped in order to make the inter-tent crossing.

The job of lashing down equipment and securing the site that evening was carried out in the teeth of a massive howling gale. As the wind picked up in its intensity with every passing minute, it seemed to us by early evening that the blizzard had achieved its massive worst. Repeatedly that first evening Ron and I struggled into our clothing and ventured outside into the feverish intensity of the howling winds to check the equipment and tents. I can remember in those early hours looking at the walls of the tent from the outside and wondering how on earth it could stand any more of this battering.

Standing up outside the tent had now become almost impossible. Visibility was less than three feet, even when wearing goggles, and absolutely nothing could be seen without their protective plastic lenses. Turning to face the wind was torture as our goggles were blown out of position by its ferocity. Our exposed position on the snowfield with the mountains some miles away was allowing the full force of the wind to roll onto us with an uninterrupted power. Even though we had placed the vehicles to the windward side of the tent to offer some little extra protection, on that first evening the tent was visibly lifting and straining against the guy ropes. As the wind roared off the mountains and struck at our shelter, the tent continued to move and lift until the build up of snow around the skirts helped it to settle into a more stable position.

Sitting on the stacked gear inside the tent, we ate our first meal to the sound of yet another blizzard, still in good spirits but knowing by now that travel the next day would be impossible. The weather was worsening, and judging by the amount of snow being blown around it would take us an age to dig out. We settled down to sit out the rest of the evening talking and turning our attention to The Voice of America at the appropriate time as the sounds were emitted from the little transistor radio.

The day, although frustrating, had just slipped by, but the weather and the paltry distance we had covered was not causing any great concern amongst us. This was traveling in Antarctica and this was what it was all about. Covering ten miles in a day's travel can be a great day, and sometimes you could even get twice that far. But by the same token, making only a few miles in a full day can be viewed as a victory of sorts if the conditions are bad. In reality traveling, no matter how slow and how frustrating, was the objective to me and I loved it with a passion.

As the evening wore on the intensity of the storm raging outside increased. Joe and I had experienced the big blow at the cape some days previously, but that had ranked as a child's game in comparison to this blizzard. We were firmly in the grip of the first prolonged winter storm as nature was ridding this land of all the detritus of the long over-inhabited months of the Antarctic summer.

By about 11:00 that evening, Ron and I were making our exit from the big tent to check the extent of any damage outside, and to crawl gleefully into the confines and the relative comfort of the small tent. Even by now with the newly born storm only a few hours old, it had begun to wreak its very own havoc, and the build up of snow on the weather side of the tent was impressive in its construction. The force of the wind had moved equipment, and even the exposed skidoos had been moved into a slightly different position.

We struggled to move in the teeth of the blizzard and point all the vehicles head on into the prevailing wind before we re-covered them with their tarpaulins as best we could. A small job that took us an hours struggle to complete. After tying the vehicles down once again and conducting a final check around, we struggled into the sleeping tent, where the next hour was spent removing our outer protective clothing and boots before snuggling down into the warmth and security of the bulky down double sleeping bags.

As we relaxed we took stock of our situation and discussed the stores and supplies we had available. We had plentiful amounts of food, much of it we could eat uncooked if the worst came to the worst and we did run out of paraffin. We could use skidoo fuel in the cooking stoves and lamps, and although dangerous, using the volatile fuel would be better than starving to death or dying of thirst. Obviously, we had enough clothing, and our survival equipment was re-sorted and stored next to each man where he was to sleep.

Our survival sacks, to be used in the event of a tent being blown away or ripped beyond usefulness, were now kept close to us constantly. Maps and compasses were stored safely and put into the weighted down box inside the large tent. At that point Ron and I were laying bets as to how long the blizzard would last. The thought of achieving my aim of two days seems tame now as the storm was to continue unrelenting in its aggressive movement for a week, during which time it did its level best to pound us into a submissive declaration of awe and worship.

That night I closed to my eyes to the pounding wind and every time I regained consciousness during the next few hours the same howling sound greeted my ears. Early, very early, shortly before the sun rose behind the clouds and tried to penetrate the barrier they created, the power of that storm began to show its true nature of destructive ability.

16
Dog Days

They dogs were all movement, straining and jerking at their restraining chains as the aircraft passed overhead, and at my first sight of them from the cramped insides of the low flying aircraft I felt suddenly I was in the heart of life in Antarctica.

From the moment I arrived at the base in Antarctica, the dogs and dog sledge travel captivated me, as they did for many others. During the evenings I would often sit quietly by the entrance to the sledging store, fifty or sixty yards from where the dogs were lined in spans, and I would wait patiently until one of them began to howl. The wolf in them encouraged them all to quickly join in a cacophony of off tune noise as one after the other they would throw their heads back, chin to the skies, while emitting a howl that would split the air and travel over the ice for miles and miles. It was a tune that sent shivers up my spine and never failed to put a smile on my face.

Thirty of these massive, furry brutes were constantly craving attention from their human counterparts, and even in the coldest and windiest weather they would jump to their feet to greet any approaching person. Shaking the hard packed windblown snow from their matted backs, they would squirm in their excitement, expectant of the run, or if not, affection. While running in front of a sledge they pulled the weight with seemingly boundless energy; when not running they would try to mug other dogs with brutal ferocity.

The preparation of getting the sledge ready to move at the beginning of a run was always the most dangerous time, as in their excitement the dogs could launch themselves into a frenzied mass of snapping teeth and streaming blood. Attempting to separate fighting dogs was usually a threatening task, especially if one of the beasts mistook the human for a target. Injuries sustained by the dog handlers were usually caused by accident, in the heat of the moment, and although such injuries never went unpunished to the perpetrator, the embarrassment and shame suffered by the offending dog on realizing its mistake caused more hurt to them than the beating they would inevitably receive from the victim. With nine dogs fighting, with flying fur and injuries, the dog handler was at his most vulnerable, and separating one pair allowed the others to continue with their violence. The time before the first run was a time of separating dogs and untangling traces.

Scientific research had long dispensed with the use of dogs as the main form of travel and we knew, even at this period of time, that the days of even retaining the dogs for pleasure were numbered. But we soldiered on in the belief that the entertainment and companionship value of the dogs, particularly to those people over winter, was immeasurable.

Full of anticipation for a good day's travel, Maurice and I set off with our dog teams across the sea ice at the first light of day. We made good progress, stopping for a hot drink to warm us after an hour or so and to rest the dogs, which had been constantly trying to pull the fully laden sledges faster and faster in their excitement. Within a short time the dogs found their rhythm, and any stragglers in the line were reprimanded and reminded of the teeth behind them as they felt the snap at their tail of less than impressed counterparts. It was all a part of the pack attitude needed to keep the momentum.

I knew well the story of a dog driver of the 1960s who had stepped from his sledge and immediately broke through the sea ice to be submerged waist deep in the freezing waters. Seriously in trouble, he had shouted to his lead dog to pull hard and the team had responded instantly and pulled the beleaguered dog driver back onto the surface and into the air temperature of minus fifteen degrees, where his wet clothing began to freeze quickly. He set up a picket for the dogs and put his tent up before removing the protective clothing around his legs, and diving into his sleeping bag where he remained for several hours. It was not simply luck that allowed him to survive; he was experienced, well equipped, and had a well trained lead dog.

We were on a short trip of no more than five hours, but both teams had full sledges, with tents and food boxes as well as radio, cooking equipment, survival gear, and skis. It was all essential equipment in case of accident or if the weather suddenly turned bad and we had to make camp.

The cold crisp clear air and the temperature of minus twenty degrees centigrade was keeping the hard working dogs cool as we were pulled along, standing on our skis by the side of the sledges. Running eagerly across the sea ice and maintaining a steady pace, the dogs gradually ate up the miles between the base and our distant landmark in the glorious conditions. Occasional shouts of command, the padding of seventy-six dog paws and the scraping of our sledge runners were the only noises we contributed to the stillness surrounding us.

Our only malcontent was one of the new season's puppies running by my side and tied directly to the sledge. The puppy was finding the going tough on his first real outing, and he was unsure of his role behind this unusual line of dogs he never seemed to be able to catch, while his efforts to pull at right angles to the sledge were being stopped by the trace attached to his brand new harness. He tried lagging behind, but found that we were not going to slow down for him as he tried desperately to sit down. After being rolled over and dragged for a distance he seemed at last to realize that being a husky was a serious business.

My lead dog, the mother of the puppy, began pulling hard to the right. I shouted, at first encouragingly, to her to steer left, and then more harshly as she seemed to be pulling harder and harder away from our target. As she constantly tried to pull away I slowed and then stopped the sledge so I could go to the front of the team to find the problem. But as I stepped from the sledge my foot sank into slush above thin sea ice, and without hesitation I got aboard again and roused the dogs, this time encouraging the leader to pull hard to the right and out of the danger area.

When we stopped a short time later, having found good sea ice once again, I anchored the sledge and approached the leader to praise her for trying to pull the sledge away from danger. Perhaps she was simply being obdurate that day. Perhaps she was just tired and wanted to get back to her home. But I like to think that she sensed the danger, either by being able to hear the seals under the sea ice much more clearly than normal, or because she could feel the difference and danger under her feet. Whatever her reason, she encouraged us to safe ground and I had been forcing her to go against her instincts. Abby was a great dog and a great leader that day.

17
Journey: Storm Second Day

Lying in the protective cocoon of my sleeping bag, I could feel the wind was now directly pummelling a different side of the tent. Ron had stirred into wakefulness. "When did that happen?" he yawned.

"I don't know," I told him. "It's just woken me up."

We both knew we had to rise quickly and get equipped to get outside. For one thing, the small two-man tent we were sleeping in was now taking the full force of the wind, and the skidoos were presenting their most bulky side to the onslaught. Struggling into our gear we began shouting through the folds of tent canvas and the screeching wind to our companions only a few feet away in the adjoining tent. Eventually we raised a response; they were okay, but there was a distinct change in the tenor of the voice answering us. The constantly raging storm had begun to burrow into our minds with its ferocity and it was having an effect on the mood.

Forcing our way outside, we surveyed the scene. Damage was limited to a few boxes but everything else was holding up in fine form. Both of us roped up, not that we feared being blown away, but more that we could become disoriented very quickly and wander off in the wrong direction. At a distance of twenty feet we would be lost, probably forever unless by luck we could stumble upon the tent.

It was much easier to crawl around the site now as the winds drove us to our knees. My hat was blown from my head and some of my long hair whipped painfully onto the only exposed part of my face. After checking all the gear and the security of the tent we forced our way through the tunnel entrance where by now hot drinks were ready for us. The blowing snow whipped excitedly around the interior of the tent as we entered through the tunnel amidst shouts of close the door and laughs as Nick dived for deeper cover in his sleeping bag. We lay back across the other bodies in the tent still immersed in their sleeping bags and tried to sort out some semblance of a comfortable position while getting out of our outer clothing.

Nick, with only one hand and his face showing from the security of his sleeping bag, was inhaling luxuriously on the last of one of his valuable cigarettes. No one else had brought any cigarettes, much to Nick's disgust, and we considered our options about what to do if the stores went completely dry. During the next few days we tried dry discarded tealeaves rolled up in toilet paper mixed with a variety of dried herbs. Little satisfaction was gained as Nick worried over the fast depleting stock and wondered how long he could resist the temptation to make them last. Even famous polar explorers had admitted that dried tea-leaves never afforded the same level of enjoyment as real tobacco.

What do you do when four large men are confined to the inside of a two-man tent for a long period of time? We read, we argued, and we passed the pee tin, the bottle used for urinating into, to empty. We drank tea occasionally and we ate even more occasionally. We slept and we discussed whatever we felt would cause some form of argument or discussion to keep our spirits up. We listened to Nick's long, rambling, and purposefully drawn out stories, and his impersonations of a supercharged Jaguar approaching a corner at Silverstone race track. As the highly tuned car roared and popped down the gears, we could visualize the flame exploding out of the exhaust pipes as the unburned gasses exploded in the sound he recreated.

We listened intently how Joe, after being released by a very famous soccer manager from his apprenticeship with a professional team, had decided to go into the world of haute cuisine. And how, soon after taking a job at a prestigious hotel, he had been delighted when this same manager had visited for a pre-match meal with his team of superstars from a major first division side. And we listened with delight as he described what he had personally done to the steaks served to the highly tuned and honed athletes and how the manager had slipped into the kitchens before departing for the game to congratulate him on his superb food.

Nick and Joe told the same stories time and again, and each time they were met with greater amounts of laughter as the wind continued with its best voice to try to drown out our enjoyment like some jealous listener trying to gain our attention.

At regular intervals of an hour or so, Ron and I took it in turns to go outside into the roaring wind to check all was okay. I usually took this opportunity to revel in the time alone amongst the tumultuous noise of that weather as I tried to look around and see if anything such as a landmark could be seen. But the visibility remained at three feet or thereabouts, and wandering off and away from the tent was an exciting and dangerous pastime I engaged in to feel the total seclusion of those days.

The story of what "Titus" Oates must have faced, as he had sacrificed himself for his companions in 1912 on the retreat from the South Pole with Scott's expedition, crowded my thoughts as I tried to imagine the type of conditions they must have faced. Although we were thousands of miles away from where that incident happened on the other side of the continent, I could intimately feel the event in all its heroic drama.

For months on the ship on our journey south from England, groups of us had argued regularly on the subject of polar travel, eventually sliding into two separate and distinct camps. One group contained those who preferred the manner in which Scott had achieved the South Pole, and the other admired the fortitude and abilities of Amundsen. Discussions like this were of little importance in reality, but I learned much from those people, who could quote verbatim passage after passage from the plethora of books written about their polar heroes. It was interesting that events of more than sixty years before were still so important and uppermost in our minds.

There was a book we kept at the base known quite simply as *The Quotes Book*. Any profound or funny quotation or statement made at the base or heard by anyone at the base was entered religiously into the pages of this book, which was always left in the main area. Some of the lines were outrageously funny, spreading back through years and years since the base had been established.

One of the base mechanics had recorded in *The Quotes Book* the story of a previous occupant of the base who, when asked what he thought of the history of polar exploration, had replied that he was having too good a time making history there himself to bother about Scott and Amundsen. Point taken. Perhaps we were still making history in the 1980s, although I certainly had never considered it that way myself. But, at that very moment one of our number, who had left aboard the ship taking the summer personnel back to UK, was well into his plans for making a massive impact on the history of polar travel. Furthermore, the journey was to be carried out in a style that Scott would have been proud of, and that would have turned the heads of many more great explorers of the past.

By the end of day two of our lie up, the weather had still not reduced its howling and raging, and as its capacity for damage increased, we settled down in the large tent for yet another day of extended chatting and reading and sleeping and laughing. Nick, having been encased in his sleeping bag for more than a day now, had still not ventured outside to relieve what must have been his now aching bowels. He was to remain in exactly this position for the whole time the gale raged. Perhaps some of us were making polar history and breaking records of a different kind.

During the radio schedule that evening with our base, as the wind screamed in the background and we waited expectantly for news from outside of our storm swept world, three or four of the personnel from the base, miles and miles to the south, opened up the channels with their rendition of "Happy Birthday." Joe produced a metal flask of brandy from yet another secret place. I was twenty-six.

18

The Boys from the Base

Around Christmas 1980, it may have been Christmas Day, several of us were in the radio shack having been told that one of our two aircraft was in the air and unable to land. The pilot was constantly circling and waiting desperately for a break in the cloud cover that had obliterated a sighting of the base from the sky.

Occasionally we would hear the distinctive sounds of the twin engines overhead while the dialogue between the pilot and the radio operator continued. Nervous, tense glances outside confirmed the worst; we could see a long way across the bay but not above fifty feet in an upward direction, and things didn't look too good.

Tension remained high as newcomers to the radio shack immediately asked, "Who's on board?" And, "How much fuel have they got left?" Many of us had made the journey to the other side of the island and seen for ourselves the wreck of the single engine plane that had been at its last resting place for many years. With the nose cone, wings, and engine removed, it remained in its awkward position in its grave, a reminder and an obstacle for the wind to bounce against and to add interest to its monotonous journeying. Although the wreck of the single engine Otter was a vital landmark for those of us who traveled regularly in that direction, few, if any of us, bothered to venture inside the remains, and few studied the records to determine what had caused the accident.

I had been aware of the potential problems of air travel in the Antarctic long before actually arriving in the continent.

A friend who had traveled and sledged with his dog team extensively in this area during the early 1960s had silenced me with his tales of travel during a period of time when there was even less backup for travelers in this vast uncontrolled continent.

The friend had told me of a flight he had been on in those earlier years. On their way to a drop off point in 1963, the weather had closed in on the aircraft during its flight. The base they had departed from was suffering from a very low cloud base, and return was out of the question and landing at their destination was also impossible. Some immediate action was called for.

Low on fuel and with a heavy payload of field equipment, including nine dogs, it was going to be difficult for the pilot to stay in the air for much longer. The human cargo were told to make their way to the rear of the aircraft and brace themselves, facing to the rear, for what now appeared was going to be a difficult landing. And the descent began.

After what seemed an age of tortuous wait as the pilot flew on blindly through very low and thick cloud, the airplane made a bumpy landing and rolled to a halt in zero visibility and no contrast. Later the pilot had explained that he simply put the aircraft into a very shallow dive and waited until his skis touched the ice. When the weather cleared they found that they had miraculously landed in a very narrow valley between two walls of mountains to right and left.

Surely now in the 1980s with all the sophisticated equipment on board, the pilot now circling the base would not have to resort to such "seat of the pants" flying. But that's exactly what he was waiting to do, and during a conversation with the radio operator the pilot suddenly exclaimed that he had seen a break in the clouds, and without pause for second thought he dived through the hole and zoomed past the base at height where the crew and passengers could almost have stepped out onto the roof of the radio shack.

The pilot managed to land at the very extremes of the snowfield on which the airstrip was marked out, and he taxied a good few miles over rough terrain to the safety of the aircraft tie downs. All in a days work. We regarded the pilots as true pilots with nerves of steel, and we felt comfortable when having to travel aboard the aircraft even over long distances without weather reports being available at drop off areas.

Disaster struck in a minor way during the end of the summer period when one of the aircraft had a particularly bumpy landing. The nose wheel was damaged, and although it did not look too bad to the unqualified amongst us, it seemed to the professionals that the damage was terminal.

I remember vividly the discussions between the pilots and the Base Leaders when the options for evacuation were being sought. They considered bringing in some sort of special pilot who would just fly the damaged aircraft back to South America; I suppose running the risk of a difficult landing when he eventually was to touch down. The proposition seemed reasonable, if a trifle dangerous, but not the option that the "powers" wanted to take. So the decision was made. The aircraft was to be taken apart piece by piece. It would then be crated and put aboard the larger of the two ships before being taken back to UK the slow way.

No one on the base at that time had taken one of these high tech things apart before, and the sea ice was going to be forming in a very short space of time. Speed was of the essence and activity was at a frenzy. Simon, our carpenter, was in his element, as it was his responsibility to supply the crates for every piece of dismantled airplane. And he made sure everyone of us was suitably employed lifting, sawing, hammering, and crating.

A Sno-cat pulled the remains of the airplane with a large heavy tracked bulldozer at the rear of the fuselage train, acting as the brake as the final steep ramp at the final stage of the journey from the airfield was negotiated. Teams of us lifted the aircraft fuselage manually onto its cradle at the shore before it was lifted onto a small boat half its size and carried out to the waiting ship. I watched with absolute incredulity as the seaman operating the crane on the ship picked up the huge red fuselage, which had been handled with tender loving care throughout the operation, and aimed it at the opening of the forward hold of the ship; an opening less than half the length of the fuselage. With a flick of his wrist the operator dropped the nose of the aircraft and it was lowered and swung into the gaping hold, followed by the rest of the fuselage, without so much as the smallest additional scratch appearing on the shiny red paint.

In a few short weeks leading up to the winter evacuation of the base I had seen and experienced first hand some of the most remarkable teamwork I had ever witnessed. Outstanding leadership and wonderful teamwork had enabled an aircraft to be taken apart and wrapped up. The men on the base had proven time and time again what incredible people they were. I was being pulled along by the skill and determination of a handful of tradesman, mountaineers, and scientists working together for a common cause.

There never appeared to be any firm orders issued by those in command, very few people lost their temper, and very few people seemed to argue. Everyone got out of bed more or less when they wanted and went to bed when they wanted. Everyone did their jobs and for the most part everything ran like clockwork.

When the last ship of the summer appeared in the bay, tons and tons of stores were manhandled onto the ice. The stores were then passed from hand to hand through a long line of heaving, smiling men and onto the waiting trailers. Transported the last few hundred yards to the base, all the stores were unloaded in the same way. This process continued for days, almost without a break, as the race kept pace with the encroaching sea ice.

A full year's supplies and backup supplies, in case evacuation was not possible the next year or the year after that, were unloaded and transported and stored in the base. It was a massive task completed in remarkably good humor, and in time to allow the ship to escape before the sea ice could enfold her and trap her for the winter.

Massive motivation and determination were the keys to keeping the bases functioning in Antarctica and these new heroes seemed to easily step into the shoes of those heroes who began polar exploration at the turn of the last century.

19
Journey: Days 3 Onwards. Lie Up

The wind did not lessen during the night between our second and third days of lie up. Both Ron and I had completed our increasingly quicker walks or crawls around the site to check the equipment. Tied onto a safety rope, we struggled in the wind against this moving but immovable object. We checked and rechecked that the lines of the tent were taught and that all the equipment housed outside was tied down and still there. It was a familiar procedure that was going to be carried out innumerable times during the next few days as our progress was to become increasingly frustrated by the severe weather.

Although the outside work only involved one of us at a time, everyone was disturbed during the hour or so leading up to departure as the thrust and shove of getting into protective clothing spread itself throughout the cramped tent interior. Outer clothing required the ability of a contortionist to get into, and the swivelling and stamping needed to get into the heavy insulated boots caused a major disturbance. The last things to put on were the goggles, and then the outer gloves, having made sure that someone was standing by to untie the tunnel exit of the tent.

A quick dive and crawl and the tunnel was closed quickly behind, allowing the minimum of hard blowing snow to gain entrance. Once outside, it usually took a few seconds to adjust to the violent change in conditions. Visibility was almost nil, and before getting into as near an upright position as the wind allowed, the process of checking the safety rope and adjusting harness position and hat position had to be carried out. The first breath was always a shock as the cold and the blowing snow poured into every opening, and even though our faces were covered with scarves, the fine white particles still found the gaps.

Return was worse as the process was reversed. Shouted commands from outside and inside would co-ordinate the opening of the tent for the minimum amount of time. In such horrific conditions, the inward dive would bring with it, especially for the occupants inside, more blowing snow, along with cold clothing and another severe drop in temperature.

Ron and I took up our regular daily positions at the bottom of the sleeping mattresses of the others. I sat at the bottom of Joe's area, Ron at the bottom of Nick's, and there we remained for the sixteen to eighteen hours of every day of our confinement. Joe made the journey outside only three times during the lie up, and Nick continued to remain cocooned in the confines of his sleeping bag for the duration. The immobility allowed us to save food and fuel, but during the times when cooking was not taking place, the temperature would plummet, and Ron and I remained wrapped in our down jackets as there was not the room to lie in sleeping bags also. Besides, in the event of a quick exit being needed, we had to remain reasonably alert.

Meal times in the evening were an event we looked forward to throughout the day. Hunger pangs constantly reminded us that although we had plentiful supplies of food, dry sledging rations did not totally satisfy the psychological need we felt for a good meal. Joe had brought with him a large jar of dried onion flakes that we opened and dropped buttered sledging biscuits into. The biscuit was then reclaimed from the inside of the jar and the taste created huge delight as the onion flakes stuck to the butter and added a variance to our diet. However, the onion flakes were strictly rationed as the storm had surpassed our expectations of length already, and that was only the third day.

The hot drinks we could afford to prepare during the long days were infrequent but eagerly anticipated. We had plentiful supplies of tea, coffee, and cocoa, along with tins of dried milk and sugar, and in addition our few tins of condensed milk boosted our energy levels, but they were not in a never-ending supply. In such a dry place it is vitally important to avoid dehydration by taking in regular fluids, but because we did not have much paraffin, melting the snow to make drinks was proving difficult.

Nick suffered the added problem of nicotine withdrawal as his dwindling cigarettes came to an abrupt end. Our experiments with tea leaves only allowed the feel of smoke being inhaled, and the revolting taste, even when dried herbs were added, only created greater problems.

Ron and I stood outside in the teeth of the howling storm in the lee of the tent with our backs hunched against the wind, and we walked ten yards or so down wind to drown out the chance of the others hearing what we had to say to each other. With our backs being pummelled by the massive force of blowing snow, with the ropes taught trying to turn us around, we stopped, and falling on to our knees we alternately leaned over to shout into each other's ears our questions and answers. We must have appeared to any passersby that we were praying monk like to appease the gods for the weather. Both of us were concerned.

The food remained, although not in abundance, still quite plentiful and we knew we could get through on cold rations. We also knew that we had plentiful supplies of skidoo fuel, although I had jettisoned more than half of ours when I depoted the damaged skidoo days before. We knew we could run the stove on skidoo fuel, and therefore could easily last for a month before beginning to worry. However, we needed to avoid using the gasoline mix for the stoves, as it could be dangerous and the last thing we needed was a fire. The tent to the traveling party is the difference between life and death, and we could not afford to lose it. But the overall position was not too bad and we could certainly hold out for a few weeks to come.

Ron and I were more concerned with the morale of the others, who had begun to sink, by day four, into longer and

longer periods of staring quietness as the severity of the
wind, increasing at times, did little to dampen their fears.
The radio schedules had revealed a continuing storm at the
base all those miles away, and travel was definitely out of
the question. Besides, disorientation in this weather meant
travel would be impossible. Our maps and compasses had
little benefit in such conditions without visible landmarks,
particularly on motorised transport, as we had to move away
from the metallic vehicles to take even the simplest reading
with our magnetic compasses.

Both Ron and I had suffered from long lie ups in the past.
Ron, a mountaineer of outstanding ability, had completed an
ascent of the North Face of the Eiger and was prepared to sit
it out in our current situation for as long as necessary. The
discomfort of the lie up for both of us was simply something we had to endure and, in a perverse way, to enjoy. We
both knew we could get through, but the condition of the
others was deteriorating quickly as the combination of the
severe cold, the immobility, lack of food, and the sound of
the extreme blowing wind and snow depleted morale. I knew
the blizzard would end sometime, but I was not sure if the
two others in the tent realized that in the teeth of such fierce
weather an end would eventually come.

We contemplated how best to keep spirits up and realized
the best way was to simply keep everyone talking and happy.
Nothing we could do would entice Nick from beneath the
cover of his sleeping bag, and we would not even try; he felt
at his most safe and comfortable in that position. But we
could do more to encourage hard and entertaining argument
and discussion.

Our conversation turned to our innermost feelings, and
Ron said he had a bad feeling somewhere that the worst was
yet to come. In the past when we had both been in extreme
conditions, we had been with people of at least equal ability.
In this situation we each had the others to consider and we
had to constantly do things for them that they felt disinclined
to do for themselves. As Ron and I were nearest the tunnel

opening of the tent during the days, we had a constant supply of pee bottles to empty between the inner and outer skins
of the tent, and although not a great chore, it began to get a
bit galling in the extreme. Now the emphasis was swinging
into that of "minders" but we were both determined to do
whatever was necessary to keep the morale as high as we
could. Neither Joe nor Nick had experienced a lie up of this
length. That, combined with the increasing severity of the
blizzard, was beginning to highlight the inadequacies of the
training system that was meant to prepare base personnel for
the rigours of the true Antarctic. Words are inadequate when
trying to describe the noise of the wind and the violent impact against the sides of the tent. The movement of the tent
can be frightening as a thin skin is all that stands between
the occupants and catastrophe. Even though Ron and I had
complete faith in the tent, we doubted the others did, and
we continued to talk and shout at each other in our kneeling position as the wind occasionally pushed us together and
apart.

When the storm worsened on day four, I thought the pre-
monition of worse to come that Ron had had come true, but
he did not acknowledge that this was the end of the difficulties. And as we huddled for greater warmth, we had to
succumb to lighting the stove once again to prepare an additional meal to take minds off the impending disaster that
could be brewing outside.

As the warmth from the stove increased the temperature
in the tent, so spirits improved, and once again we fell into
laughter and story telling. Although Joe had not moved to
assist in the meal preparation, the warmth brought him from
his cocoon like a re-emerging animal following hibernation,
and he took over the menu and pulled our makeshift meal
into another delight.

The bad feeling seemed to sneak in almost imperceptibly
on that day. I suddenly felt a change in the mood for the
worse of one of the party with the now unsmiling comments
being made reflecting an increased fear and frustration. Both

Ron and I discussed this later and felt we were being blamed for the weather, the difficulty of the cramped conditions, and the fact that we were traveling in these conditions when all others on the base were safe. Although equally dismissive of the undercurrent, we knew that we still had to get through and the weather outside suggested that it may not be for quite a while. In truth both Ron and I were enjoying the experience of this massive blizzard. We were both employed for this type of event and were experienced in this type of work. The others were not, and the faults in the system became apparent to an even greater extent.

Every year new personnel came into Antarctica to the numerous bases of different nationalities throughout the continent. We had witnessed the inexperience of many of those personnel and we had heard of the inexperience of personnel at other locations. I questioned why anyone would want to spend a winter in Antarctica with limited or no survival experience. It was true that someone could spend the winter on base without leaving the safety and relative comforts of the insulated buildings, but once in the field, off base, the situation became vastly different and the dangers increased out of all proportion.

We were not explorers, we were simply traveling in Antarctica as a kind of holiday relief during the long winter months of isolation, but nonetheless, Antarctica is an untamed place. Two members of our party had severely limited experience of the fundamentals of survival and we survived the storm, which raged with increasing power, because we pulled together as a team. We understood and identified the limitations in our traveling party, made significant allowances for those limitations, and avoided increasing the level of danger in this incredibly dangerous place.

Ron and I, as a consequence of our discussions, increased our determination to get all of us through the storm. If we had to cajole, then we would cajole, if we had to bully, then we would bully, but we would all survive this storm. And we slept side by side in the little survival tent, both enjoying the screaming wind and both enjoying the company, and continuing to make the most of this outstanding experience we had been thrust into.

Our renewed vigour during the next day kept the conversation going for most of the long hours. We caused arguments with outrageous comments, and as we knew everyone in the tent we played on each other's foibles. Nick had told me that I was from a privileged background. I had been to college from the age of eighteen when he had struggled through an apprenticeship. I did not have a trade as he did and I could not repair a broken Sno-cat engine.

Ron had similar criticism reigned upon him as he had also had a perceived privileged upbringing, going to college and spending time in the Alps pursuing his dream of climbing mountains.

Worse was to come when Nick discovered that I had hitchhiked most of my way around Africa, totally dependent, in his opinion, upon the good will of others to satisfy my desire to travel, and avoiding paying fares for my extended journey. We laughed hilariously as we came in for this heaped criticism, and the more personal the comments became, the more we revelled in the atmosphere. But the heat of the language did little, in reality, to warm the inside of the tent.

Spirits were further uplifted when Joe shouted and abused our stupidity at wanting to climb mountains for a reason that completely escaped him. But he talked with great emotion of his life and his loves and the future, and Nick contributed with his stories of his beloved home life in the small village in which he lived. As the weather continued in its powerful display of temper outside, we countered it with our fervor and noise inside.

But with the awakening of each day, the hilarity of the last was lost, and the mood at first light reflected the anger of the storm that still blew its heart out in the cold Antarctic air. The despondency caused by the continuing bad weather hit our companions like a sledge hammer each morning as they contemplated yet another twenty-four hours of confinement without decent food, warmth, or even the chance of cleaning their teeth. The atmosphere in the tent was not changing, as opening the tunnel entrance was impossible for any longer than it took to let Ron and me in or out, and the heavy insulation was trapping the fetid air inside.

As time passed things were beginning to get more desperate. It was difficult to bring everyone out of the ennui that had swallowed us. Even though the daylight was with us during the day, there was not enough to read by as the low cloud and blowing snow dimmed the sun, and the headtorch batteries and lamp fuel were too precious to waste. We escaped into our own worlds and with our own thoughts, and by the morning of the last day when only Ron was called to say a hot drink was ready by Joe, I knew we were on the downward slope and moving faster and faster in the direction of the bottom. I knew at that moment that the relationship between Joe and I had broken down. He was telling me, by not speaking to me, that I was to blame for his discomfort and his pain of not knowing what the outcome was going to be. It cut into me like a knife. I had not looked after him as I should have and his depression was turning to hate.

I had tried to train Joe over the months in the technical aspects of traveling, but I had failed to prepare him for the horrors of sheltering in a tent for long periods of time, not knowing whether the outcome would be life or death. Also, I had not prepared him for those things that can suddenly go wrong on a journey like this, such as a broken down vehicle or frostbite. There was nothing I could do in this situation as emotions can run riot, and although I understood how he felt, we had to make it back to base. If Joe was unhappy he would have to put up with it until we were safe. We would have to sort it out when we got back to base. There was nothing more we could do, except sit it out and pray for an early release from our entrapment. As the temperature outside remained constant at about minus twenty-five degrees centigrade, we would have to wait. It was too cold to escape from the tent for longer than a few minutes, as the wind chill dropped the temperature to, at the very least, minus forty. So the freedom with no escape continued.

Our diet had become tedious in the extreme, with a hot drink first thing in the morning, and some sledging biscuits with butter and jam. The biscuit, although nutritious, was rock hard and easier to take when dunked in the hot tea, and we took our time relishing the taste of the sweet drink while swallowing every morsel of the food. At lunchtime we could now only eat cold food, as we rationed the paraffin, but the evening meal was welcomed with the roar of the primus and steaming fluids as we prepared the now intolerable dehydrated meat stew.

Soon after five in the evening we looked at each other in bewilderment and relief as the noise of the wind suddenly stopped as with the flick of a switch. One minute the storm lashed us with its immeasurable fury, and the next it gave up and all went quiet. We were lying awake in the dimness of the early evening of the sixth day of our confinement when the silence shattered my thoughts. Ron was dozing but he was awake in an instant. Hurried dressing was soon followed by a quick exit from the tent and we were standing in the maelstrom of disaster of blown snow and scattered equipment that greeted us that day.

The crevasse probe used for the radio aerial, standing in the lee of the main tent, was buried up to within the top six inches. Four feet of snow had been dumped onto the safe side, the side out of the wind, of our main tent. The visibility began to improve as weather lifted, and we could see the outline of the mountains in the near distance as we walked carefully around trying to locate everything we could without tripping over it just under the surface of the snow.

Every particle of snow was hard and packed and looked good for travel; however, the blizzard had packed the snow so tightly around our equipment and vehicles it was going to prove to be a long and painstaking process to move it. So we loosened our clothing as the temperature began to rise alarmingly and took up our shovels to start the task.

20
Days Like These

From the top of the rise behind the base there was a clear 360-degree view for miles and miles in each direction. The Peninsula bordered us to one side, and the huge bay spread out to the south. The islands and mountains filled the landscape. This place was a paradise of snow and ice.

My attention from this vantage point was constantly brought back to the west, to the long ridge that led from the top of the glacier tail we called the "ramp" parallel to the track to the summer airstrip. I had been onto the ridge many times, and had climbed several of the gullies and the main buttress, and I knew the conditions on top and the state of the rock and snow and ice. With binoculars I now viewed the knife-edge ridge and the faces and gullies, and decided that it would be a tremendous prospect to traverse the complete ridge from the airstrip some miles off to the finish on the top of the face overlooking the base. On the way back we would probably need to rappel down a gully only one hundred yards from the very end of the ridge. A long and very steep, claustrophobic gully recognized by us all because of the generators mechanic's attempts at getting up its ice-encased surface earlier that year.

Once the plan was discussed with Wilfred, we recruited Maurice and a few of the others to climb with us for the prospect of a good day out. The route was not difficult and would mainly entail some good scrambling in roped pairs. We would even be able to move together most of the time, so the restrictions on time caused by the shortening of the days in the lead up to winter would not cause a problem.

It would be a good route to do on a Saturday, as we could finish the day with the normal and formal evening meal held every week. So in the very early hours of the day in the earliest part of winter, we bribed and cajoled Nick, the tractor mechanic, to get out of bed early, forgoing his Saturday morning lie in to take us all by Sno-cat to the far end of our route for the start.

The weather was superb, with seemingly limitless visibility and hardly a cloud in the sky with a temperature of minus-ten degrees centigrade. A strong, bright sun beat down upon us as we began the long hard climb to gain the ridge, and as the big Sno-cat and driver disappeared into the distance we experienced the total stillness and quiet of the early day.

On reaching the top of the first climb, we had an uninterrupted view of the ridge stretching out and tracing its way to our destination. The ridge was knife edged for much of its length, with the spread of snowfields to the west and the draw of miles and miles of ice and cornices. For hours we climbed and descended, moving quickly between the numerous small peaks that formed in a line along the ridge. Encountering only a couple of tricky moves, we sought out difficulties to add to the enjoyment and the amazing views from the knife-edge.

Our first break that day was taken sitting just below the ridge top on a small ledge gazing down hundreds of feet to the snowfield below. We could trace our footprints left in the snow in the distance as the ridge spread back towards the airstrip, and we were impressed with the speed at which we had been traveling. Food was supplemented with tea and condensed milk, and we languished in the bright sunlight, not really wanting to disturb the moment by the inevitable need to move on. But we still had a long way to go, and the noise of camera shutters capturing the moment and karabiners snapping onto ropes signaled the beginning of the rest of the day on this tremendous piece of mountain.

Skirting low down the side of a cornice, the wind blown snow on top of an edge, we looked up at the incredibly steep and uninterrupted sweep of snow and ice, which was as smooth as cream and as white as could be. I needed to get to the top just so I could look over and down into the valley beyond, and Maurice, Wilfred, and I slogged upwards in pristine conditions for three or four hundred feet. With the grace of three characters from an Ealing Comedy, we gingerly peered over the edge into the massive panorama of snowfield beyond. The view was one that was too good to compare with anything I had seen before. Without signs of any other tracks there was no evidence that the view from this position had ever been seen before.

Quickly rappelling, we joined the others before moving on, and the joyous view over the cornice was lost until it could be recounted that evening for the first time in the usual exaggeration of the climber's day.

Hours later we peered out over our last peak of the day onto the sea ice of the bay, perhaps a mile from the base to our left which was yet unseen. An extended drink in the lee of the quickening wind and dropping sun, along with the joy of the last cigarette of the day on the mountain, and we were ready to rappel down the gully and onto the track back to base. Forty-five minutes of walking through the rapidly increasing twilight, with the lights of the base twinkling in the cold air, we arrived at the base just at the time when we would have had to switch on our torches. The twelve hours had been just enough time for us to complete this great day.

Throughout my years of mountaineering I had experienced some unforgettable times. Some in the high mountains of the Alps had been truly magnificent with hard climbing, long days, difficult weather, and superb companionship. There had been other days that had been magnificent, not for the severity of the climbing or for the triumph the climb had allowed me; they were magnificent for other more esoteric reasons. One of those days, very early in my mountaineering life years before, had been in the mountains of North Wales in the UK one Boxing Day. The company of friends on that day and the conditions had been sublime, and huddled into the lee of the grossly out of place building at the top of the mountain, we had enjoyed a belated Christmas meal as the wind and the snow buffeted us from all directions. As the sun lowered we all charged from the mountain-top, cutting straight down, and we descended into the valley in record time. That one day will remain with me forever as a truly great day. It was, unfortunately, the last I was to spend with one of those friends, who was to perish soon afterwards in an avalanche only a few miles from where we had climbed that Christmas.

The day on the ridge in Antarctica was also a truly great day. Not simply for the climbing, but for the real meaning for me of mountaineering; the sheer enjoyment, the outstanding companionship, and the opportunity of becoming more at one, and at peace, with the mountains.

21
Journey: Too High

The devastation of the blizzard can only be adequately described by the amount of time it took Ron and me to dig out all the gear from underneath the blown snow. We attacked the hard packed deluge long before the first light of the day was thrown over us and worked unceasingly, digging, wrenching, and pulling at the submerged gear scattered around the site, for four hard hours. At midday, we sat with the others to have our last meal in this place, exhausted and wringing wet by the effort. Morale was low as we faced each other in the tent, but the warmth of the food encouraged us and we regained our eagerness to move.

Nick had again stripped the carburettors from the skidoos and cleaned them, the last act in the erected tent, and the mounds of remaining gear were loaded and secured on the sledges. The efforts of the morning gave way to our departure at 1:15 p.m.; it was the last lap.

During mid-morning, a white out had gradually begun to enclose us, and although it had not totally cleared when we started our engines, we ploughed onwards. Ron and I discussed the conditions, and although not totally happy, we felt confident with the weather. Our skidoo train was strung out over a distance of about one hundred and fifty feet, in line astern, with the straightness of the line being confirmed by the rear man. The weather, keeping to its game, kept changing and we kept stopping. In the bad visibility we took advantage of the delays and refuelled the skidoos, and drank the hot tea from our flasks. There was an eeriness on the landscape that was exaggerated by the occasional poor visibility and the stillness and quiet of the day, but the magnificence remained even though it seems, all these years later, that the place was reluctant to let us leave.

I was concentrating hard on making sure that my skidoo kept directly in the tracks of the two in front and that their line did not wander, and my engine strained to keep the momentum going. The bright orange of Nick's anorak immediately in front, and the blue of Ron's, filled my vision as they tried to huddle low on their machines against the cold of our continued progress. Joe, behind me at the back of the sledge, was suffering in the cold. Although I could not see his face I knew he was not smiling, and he stared ahead, rocking and rolling with the movement of terrain. It was strange that my engine was straining and its noise kept bothering me. It shouldn't have been straining. With a shock I realized we were heading upwards and we must be off course.

I stopped and kept the engine ticking while I looked around to make sure I was right. In the murky gloom, the mountains to our right were too close. "Shit!" I exclaimed loudly to no one but me. I was sure of it. We were too high. A quick glance at my watch showed we had been traveling for less than two and a half hours. We must have turned onto the glacier too late, past the point where we should have begun to head down onto the troubled icefall. We hadn't been climbing too steeply but the straining engine had been trying to tell me we were traveling upwards instead of down.

Keeping control of your traveling line in a white out is hard. The natural line that is the horizon disappears and the land and the sky join into one continuous white mass with no visibly defining break. A blank white canvass is presented to the traveller, and keeping focus is difficult and really only possible if you have something solid and dark in your sights. The dangers increase as the contrast, needed to distinguish troughs and holes, is no longer there to help. Even though we were not in total white out when I discovered our course error, it was still very hard.

During a journey months previously, I had been leading another colleague up another glacier. I was standing on my skidoo footboard and driving the machine hard, keeping my focus intently on the apparent progress of my machine. An orange shape appeared on my left and my colleague dismounted from his machine and walked the few feet separating us while I thought I continued to plough forward. As he leant against the plastic cover of the fairing of the machine, I realized that I had been in the same spot making no forward movement for some thirty seconds. On another occasion on the plateau, a whiteout descended quickly and I decided to stop our progress as I was on unknown ground, and although I knew there was a mountain range to my right, there was no sign of it. The weather was warm on that occasion, and we lounged in our seats drinking coffee when just as quickly the visibility returned and I viewed with horror a snow ridge some quarter of a mile directly in front of us. There was a flat approach to the end of the ridge with a drop of some one hundred vertical feet over the other side. Although we were not off our course, my skidoo was pointing directly at the snow ridge.

The difficulties of checking route and progress in white out mean it's usually the best policy in such conditions to stop. Following a wait to hope the weather clears it is good policy to get the tent up to regain some heat. I had been caught, many times, in the position of taking the tent straight down again as the powers in the sky joked and lifted the weather problem as soon as shelter was erected. It could happen many times in a short day, but if there was plenty of light left to travel and the weather improved, distance could still be made.

This time the visibility, although not perfect, was good enough, and we had allowed the depleted morale and increasing problems to dictate our decision to travel in the less than perfect conditions. However, we knew the weather would lift and the progress need not be halted. Besides, we all knew the route well.

But now we were too high. Now we could be in trouble.

I opened the throttle and powered after the others, quickly overhauling the second sledge in line. Drawing level with the first skidoo, I motioned frantically to Ron, the driver, to stop and cut his engine.

Before stepping down I probed the immediate area for crevasse dangers and then made towards the other vehicle. Not interested in a long discussion, I made myself clear as the weather began timidly to lift and it became apparent that we were off course. The line of the correct track, although not yet totally visible, was a few hundred feet below us, deep down amongst the low cloud. The decision was made and I would start the progress back to intercept the correct track as the weather began to lift more and the way became visible. We did not need to rope the vehicles together to brake the speed as we traveled very gradually down hill. The ground we were pointed towards was all undisturbed and should be safe to bear the weight of skidoos, sledges, and men. Joe was collected from the rear of my sledge and he eagerly joined Ron on his skidoo, flapping his arms to regain some body heat. I confronted Ron and told him Joe should ride on the rear of his sledge. But Ron thought he would be warmer on the skidoo and that it would be okay. At that point we had made yet another mistake.

I probed in front of my skidoo before getting back on board, and I stared into the distance, looking for the tell tale signs of sagging, weakened bridges. The slightly darker color in the snow was a sure sign of a depression, possibly meaning a crevasse bridge could have dropped slightly. But the immediate ground looked clear, and I started my skidoo into life and began the cautious and slow gradual descent towards the route.

Less than fifteen minutes later my skidoo uncontrollably pitched to the left, fell onto its side and rolled downhill. It had been pulled over by the sledge, which had been sliding downhill and pulling at the towline. The roll threw me off the skidoo and instinctively, as I hit the ground, I rolled away from the weight of the out of control skidoo, and pulled myself to the length of the safety rope connecting me to the surging mound of metal and tracks. As I fell, the automatic cut-off line attached to my right wrist pulled out and the skidoo engine stopped, but I was shocked as the skidoo and I continued to roll. The gradient was too steep.

My downward progress seemed to continue for at least three or four complete rolls, with the vehicle coming to a standstill within feet of where I lay. Below me the overturned sledge had also come to a sudden stop as its cargo dug into softer snow. I lay on my back in the snow as I checked for any injuries. But this had been a minor fall, and as I began to rise to my feet, I heard a shout. I pulled my goggles from my face and looked up towards the second skidoo with the driver and passenger. But there was nothing there. I did not understand. There was nothing where the skidoo, sledge, and passengers should have been. There was only the now bright sky and the outlined moon nestling in a dip between two unnamed peaks, above the line in the background. Two unnamed peaks towering above me seemed alarmingly close to where I was now kneeling, and I moved my eyes to the right where Nick was visible and sitting motionless, still aboard the third skidoo, looking straight ahead.

Confused, wondering what the hell they were playing at, I unclipped from my safety rope and heaved and thrust myself uphill, and from somewhere the word crevasse forced its way into my thoughts. I paused briefly before clambering on. I still had the white moon with its powder blue patterns filling my vision, and I could hear my breathing and the sound of my feet breaking through the top crust of the snow. Everywhere else seemed to be flooded in silence.

Nick had walked forward from his skidoo before I shouted at him to stop and go back. "They've gone!" was all he said, as I first looked at Nick and then reluctantly at a small dark blue hole in the snow; a hole that was no more than three feet wide, down which men and equipment had disappeared.

Think. *What do I do now? Make the area safe. Check the victims.*

I screamed at Nick, "Don't move. Stay aboard . . . don't fucking move." *Let me think let me think.* "Stay there. Don't move." If the crevasse was under me now I was in trouble. I was un-roped fifteen or twenty feet away from the relative safety of the third skidoo, so I lay down to spread my weight more evenly and inched carefully towards the lip of the gaping hole. It was no good. I couldn't get close enough, but by now I was more or less certain that the crevasse was running diagonally away from me. But I was unsure of the strength of what remained of the damaged bridge.

Still lying down, still facing the crevasse, I edged back feet first in my tracks towards the upright skidoo with Nick aboard. With my feet almost in contact with the front ski of the skidoo, I jumped up, grabbed the safety sack stashed behind the driver, pulled it open, and uncoiled a rope, which I tied to the skidoo. I then secured myself to the rope with an Italian hitch to the karabiner and blue sling that ran through the loops of my harness and through the crotch karabiner. All completed in a blur of movement practiced time after time after time. Procedures I could carry out with my eyes closed and my hands enclosed in the massive weather protective bear paw mitts.

I was now safely attached to the skidoo and I could move away in control by keeping the rope tight. The friction of the rope running against itself through the knot at the karabiner provided the security I needed as I probed around the remaining skidoo to check the ground was solid and that we were not sitting on another or the same crevasse. Nick had done exactly as I ordered and remained motionless—he had not moved a muscle, not even to waste time talking. I did not even consider at this time that he had already probably gone into shock.

Turning to face the hole in the crevasse bridge, I slowly played out the rope with one hand and probed for weaknesses in the ground with the crevasse probe. My equipment for self-rescue was attached to my harness, so I dropped the probe and lay down, crawling to the edge of the hole. At least if I fell through the bridge into the hole my jumars would get me up the rope. With the rope tight and the Italian hitch controlled by my right hand, I managed to peer into the dark blue hole. The blue color became deeper and deeper as the crevasse went deeper and deeper. Blue turning to black as the crevasse dropped into the unknown depths and my world plummeted with the realization that the crevasse was massive. Not only deep but also wide; surely, too deep and too wide. Surely it was too wide to have been bridged. I peered into the depths, unable to see any sign of the sledge, skidoo, or the passengers.

"Okay, okay don't panic; they could be under the bridge! How?" I asked myself. "'The crevasse could have narrowed quickly; it may only be a few feet deep, a trough in the snowfield." I was unconvinced and did not reply to my nonsensical questioning.

The crevasse must have been about ten feet wide at its top. I could only guess this, as I knew I was well onto the bridge and my life was supported by four or five feet of fractured, frozen water. I could not see under the bridge, but the line of fracture was apparent and the bridge had given way at one small point. The marks in the snow at the other side of the fracture meant that the skidoo probably cleared the bridge, but the weight of the passenger and driver must have pulled it back and down into the abyss, immediately dragging the huge sledge with it. I had traveled over this very spot in my lead skidoo and I may have weakened the complete bridge structure with the weight of my vehicle and sledge.

I lay on the bridge at the lip with my head hanging into the hole, staring into the nothingness beneath me. I shouted, waited, shouted again, waited, shouted again.

22
Everyday Life on Base

With the onset of the winter season, everyone on base had settled into a routine of working at our jobs. For me the smells of the sledging store seemed to be my constant companion for long periods as I repaired broken sledges from the season before with the glue dope used to shrink the twine lashings on the willow frames. Hours and hours of time were spent amongst mounds of gear. I unpacked and sorted through the remains of the survival sacks from the season that had just finished. All the old and damaged equipment was replaced with new gear. Sleeping bags adorned every available hanging space as their down fillings were aired, and the tents were systematically erected in the store to be checked, repaired, aired, and repacked for the journeys to come in which they would play a massive part.

Every one of the many safety harnesses were carefully checked for abrasions before being hung up in some kind of renewed order, and crampons and ice axes were sharpened and cleaned. Miles and miles of ropes were rescued from discarded sacks and sledging boxes and checked, cleaned, and hung to dry, and the gear our winter traveling parties was to use was checked and packed, ready at a moment's notice for the journeys. Amongst all this preparation and work the day-to-day life continued. Dogs had to be fed more regularly now as the temperature dropped. The seal quota for feeding the dogs had to be achieved, and the results of the hunting had to be dragged on to the seal pile where they immediately froze in the open air, deep freezer.

Every day someone was required to saw blocks of the hard packed snow from around the buildings for depositing into the snow shoot to the melt tank for our water supply. It was a laborious but vital task to keep the water supply at the required level, and it enabled all of us to indulge in the luxury of our weekly shower as well as providing our daily water requirements.

Blown snow, which began covering the buildings, was moved around, and the chemical toilets had to be emptied on a daily basis. Barrels of skidoo fuel were stored and the massive dump of aviation fuel stored in barrels had to be transported to a safer and more sheltered place in order that it could be located in the aftermath of the winter weather.

As we poured our energies into the varying chores that had to be completed, breaks would be taken regularly in the dining room. Most of the personnel would sit and talk as we drank tea and ate at every opportunity to provide us with the energy we so needed in the increasing cold. The only person missing from these gatherings would be the night watchman who, after the solitude of the night, would be sleeping.

Ironically, even with all this snow piled high around us, we lived in a very dry environment, and the dryness created the threat of fire, one of the most absorbing problems we had to face. Wood became tinderbox dry within a very short space of time, and as our buildings were wooden, the night watchman's job was that of fire watch as well as that of meteorological observer and recorder.

Inevitably, talk amongst us not only centered on the next journey or immediate problems or what the weekly film was going to be, but how and by what method we would travel back to the UK when our time came around. Many would forsake the ship trip for a more thrilling or a more relaxing return. Maurice and I had decided we would travel back through South America and take in some of the climbs on the remote mountains of Aconcagua and Alpamayo. We would also spend as much time as possible making our way north, and even though we were twelve months from the time of departure we began our preparations.

On hearing of our intentions, Wilfred, our resident expert in Spanish, gladly offered his services of tuition in the language. Gratefully received, we prepared ourselves for the first lesson in our library and were met by another four of five people as we arrived late. Amidst hilarity we realized that our teacher had taken on the task with massive enthusiasm and a level of seriousness we were quite unprepared for. Wilfred had provided us with books, pencils, nouns, and adjectives, and a structure to the lessons that became frightening in intensity. Homework became a regular feature as we battled through tenses and subjunctive verbs, and the lessons would start later and later as we errant schoolboys tried to finish our assignments in the last few minutes before assembling for another torturous lesson. The teacher chided us for our lack of concentration and misplaced books, and toilet breaks were banned as lame excuses in this "Dotheboys Hall" of the south. During one frightening lesson we were informed that the following week we would have a test, which we would have to complete under correct examination conditions. It was too much to bear, and the cracks that had been showing in the intense academic environment for some time began to widen. Letters from our mothers begging our absence had no impact on the stern teacher as he prepared the test questions. Gaining advance knowledge of the questions became our aim, and raids were planned and carried out on the teacher's room. But, no matter how clever we were in the build up, Wilfred was cleverer, and as the day dawned he physically sought every one of us from our hiding places and dragged us into the examination room.

The results were a catastrophe for me. "*Donde estre la oficiano de correos per favor?*" was my answer to practically every question, and the result went down as a miserable failure in my report. Maurice faired little better, and although he had kept repeating "*Mi no habla Espanol,*" neither of us relished the thought of what it would be like facing our parents that night with our failure. The top of the class barely achieved ten per cent of the marks, and Wilfred felt that he was banging his head against a very hard wall. Reluctantly he abandoned his efforts and retreated to his stores, where he was preparing the tents for another season.

It was during one morning's work that I found Wilfred lying on the floor with the top half of his body buried in the folds of a collapsed brand new tent. "Are you alright?" I asked, to which he stuck his grinning face out from his partial hiding place and said, "Quick come in before it goes." I likewise lay on the floor, wondering what I was going to see, and buried myself in the folds of the brand new tent before Wilfred said, "Can you smell it?" Then it hit me; the massive smell of perfume pervading the folds of the material. Some unknown female had packed the tent back in the UK, and we spent joyful minutes breathing in the luxurious fragrance lying on the floor of the sledging store, reveling in our imaginations. It had been a very long time since we had seen a woman.

The radio operator, helped me with the secrets of operating the short wave radio, and I was granted a rare Victor Papa call sign. During one afternoon each week, I was able to transmit, and callers throughout the world would try to log a contact. The activity kept me entertained for a few hours each week, particularly if a female voice was heard, but there were so many other more interesting things to do I missed my slot more often than not.

One very cold and very clear day Simon, Wilfred, Maurice, and I tried our hand at Inuit style fishing. On the sea ice at a distance of about fifty yards from the shore, amongst the tangled remains of small icebergs, we drilled holes through the three feet deep, frozen sea and dropped hand-held lines into the depths below. With hooks baited with bits of bread, corned beef, and anything we thought would entice some of the ice fish that were apparently in abundance this far south, we settled back on sheepskin matting and waited. And we waited and we waited until an hour or two later with a shock my line went taught.

After allowing the fish to nibble for a while, I struck as hard as I had been taught during my younger days of fishing off the coast in the North East of England and the rivers of Northumberland. The fish I pulled to the surface was not like anything I had caught before with a very large head and a small body. The creature put up little fight and I grounded my first Notothenia. It was met with rounds of derision by my fellow anglers, but I was determined to keep my prize that froze rock solid as soon as it was exposed to the cold air. Within minutes I struck again and pulled up through the hole the twin of my first catch, quickly followed by its triplet, as my fellow anglers were becoming obviously frustrated by my success. It had become a competition and bait was stolen, new holes drilled, and lines reset as we fought for the best catch.

After five self-gratifying catches to my friends' collective none, I decided to get back to the base to get this luxury of fresh fish to the cook. Delirious in my success, I pulled away on the skidoo as the others began to pull in their catch, and by the time I got back and alerted the cook, the others returned and our now larger collective catch was pooled and prepared. One of the fish hardly made a meal, but we sat at the tables amongst the other base members resplendent in waders and fisherman clothing, eating our catch, and, in true fisherman style, yarning about the ones that got away.

An old, largely hand written book in the base library detailed some of the delicacies enjoyed by explorers during the early part of the century. In our situation, food was never a problem, as we had copious amounts and lots of variety, but in the early days of polar exploration, food was very often a problem. It is well documented that even when food was hard to come by, seal blubber would rarely be consumed. However, Shackleton demonstrated, in the severity of the situation in which his expedition found themselves in 1915, that the forlorn explorers would eat anything they could. During the famous retreat from the sea ice of the Weddel Sea, his men would eat the inches thick layer of blubber fat, boiled, fried, and even raw. In 1981 we never felt inclined to follow suit or feast on cormorant or skua, and certainly we were not going to catch elegant penguin to serve as our Sunday roast.

Early one Sunday morning, deep in the winter, a solitary Emperor penguin appeared. It proudly stood around for a few hours, watching and enjoying the coming and going of these strange creatures in this stranger environment. At three and a half feet tall, it was a mature bird with the most beautiful and iridescent plumage that it occasionally preened to show its disinterest in us. One of the biologists on the base informed us that the nearest rookery for these magnificent creatures was more than eighty miles away, and presumably the bird had become lost in one of the frequent storms we were now experiencing. We wondered how it would find its way home, but we need not have worried, as during the afternoon, satisfied that it had seen all there was to see, it turned its back on us, flopped onto its stomach, and tobogganed into the distance over the sea ice.

The next morning there were two Emperor penguins waiting for their first sight of a human, and the following day several more as with the day after. They were fascinating to watch in their natural habitat and obviously just as fascinated with us. The Emperor penguins provided a distraction from the more usual Adelie penguins that waddled and raced across the face of the base, trying to avoid as much contact with us and chastising the spanned out Husky's as they passed close by.

The short sighted Adelie penguins would catch sight of one of the marker barrels around the base and would waddle towards it thinking the black outline was a penguin. Once they had established that the barrel was just a barrel and not a penguin, they would catch sight of another barrel in the distance, and so the process would begin all over again. Eventually they would stumble onto the sea ice and disappear into the distance.

As the winter snows built up to their twenty feet depth around the windward side of the buildings, tunnels connecting the carpenters shed and the puppy pens were fashioned.

The electrician moved tons of snow in the heavy bulldozer in their construction, and the finishing touches were added with shovels as the thirty feet long grotto was given its finishing touches. It meant that Simon, our resident carpenter, could wander over to his workshop and create his splendid work while the everyday movement of vehicles and men could continue on the snow above.

Simon was one of the stars of my Antarctic winter. Lanky and longhaired, he looked the archetypical hippy. Laid back in the extreme, he had an attitude to life that was refreshing and conducive to his trade in this wilderness. He would wax lyrical about a piece of wood, pointing out grain lines and relishing the texture before he worked upon it, spending hours taunting and teasing the wood into whatever needed to be made. He would then proudly stand back as we, in turn, would wax lyrical about the astonishing results he achieved.

Simon entertained us for hours with long, convoluted stories that often degenerated into a mash of nothingness if he was drinking beer. Within a short time he would become unintelligible as he wrestled with sentences, increasingly affected by small amounts of alcohol, but he always played a brilliant hand of bridge no matter how hard he had to struggle with his slurring words.

Occasionally, Simon would venture out to join us on the ice cliffs and he quickly learned rescue techniques. Unfortunately, I only journeyed with him once or twice in the early winter, but he proved to be a great companion on the ice, moving carefully during the days with increasing confidence. Simon loved the environment and marveled at the amazing scenery and quickly changing weather patterns as he offered this part of his life to his personal exploration. In the tent, as we camped on glacier and ice plateau, he entertained me with his enthusiasm for life and stories of his beloved Coventry AFC soccer team that he had supported passionately through the years. When on base, Simon mastered his dog team with great determination, and I enjoyed enormously being invited by him to be a passenger and extra weight on numerous occasions as he drove his dogs. We would travel miles onto the sea ice as he coaxed and cajoled his team in the dense quietness of those days. He would intersperse his Inuit commands with loud swearing if the dogs did not respond immediately, and his team obviously learned his style as they all furtively glanced behind them when the expletives traveled to them across the cold air.

It was hard to believe. There we were, two totally different people, enjoying immensely the surreal experience of driving huskies through the snows of Antarctica. Simon had spent most of his working life on building sites throughout the UK, with tools in his bag, wondering about things in the world of construction. Now he sought his own freedom and excitement as we shared the experience that was the barrenness of this un-littered and proud landscape, and we both found, in our own ways, what we had been looking for throughout our years.

In my thoughts, I look forward from my seat astride that bucking, twisting sledge, and see the forceful dogs pulling with all their combined might as we were dragged up short rises and around grounded, majestic ice bergs locked in their place on the frozen sea by the winter. The rear pair of dogs in the line were brothers. With a combined weight of two hundred and forty pounds and enormous strength, they would bite at the reluctant heels of the less strong dogs immediately in front of them. Nine dogs stretching away in front, being led by the smartest of them, all with tails raised and curled, tongues lolling to the sound of their padding feet and breathing heavily as they gasped in lung fulls of cold, crisp air in their immense effort.

Now, I can recall with great affection the noise of those sledge runners scraping over the ice, the pure joy of the run, the freedom, the friendship, and the sheer awe with which we viewed that inspiring place.

23

Journey: Crevasse

There was no response. I must have shouted twenty times but there was no response until, quietly, from the deep blue gloom of the depths of the crevasse, I heard a tired cry.

Moving carefully, I retraced my crawl to the remaining skidoo and regained my feet. Nick was still astride the machine and still looking straight ahead, waiting for me to tell him what I was going to do. He was still tied to his safety rope, so I extended it with another climbing rope and we retreated carefully to the remaining sledge.

"Ron's alive." I said. "One of us is going to have to go down, and I think it should be you." I offered no reason.

Nick had never been in anything like this situation, but I knew he could rappel and jumar, and I knew that if there were to be any more trouble I could pull him out of the crevasse. I was not certain that he could have done the same for me.

"No way," he replied. "I couldn't do it. You've got to do it, it's your job."

He was absolutely right of course. It was my job; I was the one with the skills to do this. I would have to rappel down into the crevasse and try to get Ron and Joe out. But in the meantime, I needed to get Nick onto safe ground to operate the radio.

We unpacked the radio and set it up with a short aerial. My colleague was tied and secure, and although the temperature had begun to fall alarmingly, I could not waste time putting up the tent. Nick had to huddle down out of the wind while he started sending an SOS to the base. I think at this time it must have been around 4:00 p.m. and although there should have been a constant radio watch, we would be lucky to raise anyone, particularly as it was Saturday.

As I was preparing hurriedly to get into the crevasse, I could hear Nick, even at this early stage, beginning to get irritated with the nothingness coming back to him from his repeated appeals for help. He continued unrelenting for hours.

Ten feet from the lip of the crevasse, as I sorted ropes and equipment, I felt encumbered by my huge Russian wolf-skin bear paws; my progress was too slow. With a flick of each wrist the bear paws dropped into the snow on either side of me. Discarding this vital equipment was to cause me severe problems as I cut and battered my freezing fingers against the ice of the crevasse wall. In the years that followed and during my flashbacks, the insignificant movement of dropping my gloves played havoc with me as I relived the simple action time and again. I can still feel the cord that joined the bear paws together around my neck flick over the collar of my insulated jacket, and I can remember the cold on my hands as they were exposed to the elements. I can see the knot in the cord joining the bear paws together, and I can feel the warmth and the comfort they had provided. But most of all I can still see the bear paws lying at the lip of the crevasse where they fell, and I have tortured myself repeatedly, over the years, with the question of why I discarded them when they were attached and hanging around my neck. I do not know why I did not retrieve them from the scene later.

With my rappel device attached to the rope, I edged towards the lip of the crevasse again. My weight on the bridge, already damaged by the collapse, caused further fractures and crumbling as I put my feet onto its outer edges. The first stage of my descent was accompanied by the "whoomph" of more collapse as I fell through and dropped a few feet before the wrench of my rope brought me to a stop.

Shaking the snow out of my hair, I looked to left and right to survey the extent of this monster before peering down between my feet into the nothingness below. The two spare ropes slung over my shoulders were going to be needed as I shouted and rappelled.

Forty feet of descent later, I heard a response. My name was being shouted weakly and repeatedly now as I wondered how anyone could have survived this fall. Disbelievingly, I came to my first major problem—the rope was too short. I looked down past my boots to the figure of eight knot in the end of the rappel rope flicking around wildly only a few feet below me. Countless people have rappelled straight off the end a rope because they have failed to put a knot in the end of it—at least I had done one thing correctly.

I halted my slide and pulled up the rope end before knotting it on to the end of one of the spare ropes around my shoulders. Dropping the now extended rope downwards, I needed to secure my weight on the rope above the knot so I could move my rappel device from above to below the joining knot in the two ropes.

Crevasses do not get this deep surely, but I could just make out way below me the tangled mess that I was later to learn was the skidoo, wedged like a cork in a bottle against the crevasse walls.

More of the bridge collapsed in on me, the snow pouring down the neck of my jacket. But with the collapse more light filtered down, and I was able to make out the person who was calling to me. His position seemed wrong. Another collapse, more light, and more of the carnage became visible.

With the ropes now joined, I clipped into my jumar, above me on the top rope, and took my weight as I released and repositioned my rappel device for the continued descent. *Retrieve the jumar, clip it in to my harness, and get going.* Within seconds I was standing on the jammed skidoo, looking around in total disbelief that anyone could still be alive after such a fall and amongst all this wreckage.

I couldn't take it in as I tried to avoid the thought of facing the injuries that I was now amongst. Ron was jammed between the skidoo and the crevasse wall at his lower left leg.

It was almost completely severed and I could here my voice involuntarily saying to him, "Your leg's broken." He had a massive head wound which had left streaks of almost black blood in a line down the wall of the crevasse, originating probably from the point of impact some ten feet above where he now lay. His right leg was contorted in an unbelievable way, and his femur was almost certainly broken, as was his pelvis, and probably much of his lower leg was in the same state. His breathing was labored, and he frothed at his mouth as he gasped in the freezing air to feed his so obviously damaged lungs. But he wasn't ready to die at this stage.

I talked to him non-stop as he passed through a period of intense pain, which just as quickly as it had started, simply stopped and his lucidity returned. His technical expertise shone through the dim light as he tried to help me.

But I had to think. How the hell was I going to move the skidoo to get him out? How was I going to treat his injuries? I didn't even have any decent painkillers to give him. I hated the man who decided to remove the morphine from our first aid packs, which had until then been affectionately known as suicide packs. But, I thank God every waking day I have that I didn't have morphine, as it was obvious that there was only one outcome for this horrific frenzy taking place. In my wildest dreams I do not think I could have put my friend out of his contorted, pain racked misery.

I lay across the upturned skidoo and leaned over the rear to look underneath, but I was totally unprepared for the sight. Joe was there. Hanging quietly in his harness only a few feet from my face. His body was wracked and torn with the aggressiveness of the fall. Restrained by the safety rope he had been kept close to the fierce metallic frame of the skidoo as it shot downwards and subjected my friend to as much damage as it could inflict. I was appalled. The damage and the blood overshadowed any minute flicker of hope for life; the candle in that second had been snuffed out.

Even so, I leaned down and over, and pulled the inert body towards me to check for signs of life, praying that none existed. He was dead. Joe, my traveling partner, was dead. I gently released the rope and as Joe swung back into the position in which I had found him I realized his beloved black hat that he wore constantly was finally gone. My attempts to hide his hat over the previous months had reached a final conclusion and we would never see it again. As I continued to stare for what seemed an eternity, unknowing the sight burned itself into my soul with a hostile permanence. Had I known then what I know now, I would not have lingered, and perhaps I could have protected myself from the ravages of the long years that have followed as the memory haunted me wherever I was and whatever I was doing.

The weight of the sledge, hammering down the hole, must have snapped the tow rope that attached it to the skidoo with the massive impact, and it would have plummeted into the depths below. Although now, all these long years later, I am not certain, as I remember a dark mass further along the crevasse, which could have meant the sledge was also wedged in place. But I am sure there was no sign of the towrope that should have been attached to the rear of the skidoo.

I pulled myself back up onto the upturned skidoo and faced Ron. "How you doing?" I said rather without conviction.

"Is he dead?" was his response, and he knew from my silence that he was. "You're going to have to cut Joe free to reduce the weight," Ron said to me, and I reached into my pocket and took out my brown lock-knife with the damaged handle. I opened the knife, hesitated and closed it and replaced it in my pocket where it had come from, without using its razor sharp blade. I simply could not cut the rope and let the body of our friend Joe fall free into the bottomless depths.

The life was draining from Ron, but I could not give in; after all, he hadn't. So rather pathetically, I spent some minutes chipping away at the iron hard ice of the crevasse wall close to his trapped left leg. I thought if only I could move the skidoo I could get him out. He wouldn't fall any further,

as I had attached him to my rappel rope. He had been traveling on his skidoo unroped—not that it mattered now or even
then. Strangely he had stayed in contact with the machine
on its fall. He hadn't even had time to jump off when or if
he had felt the machine going down through the fractured
bridge.

My continued chipping hardly produced a mark on the
wall of the crevasse. My tools were inadequate.

After a while I stopped my pointless activity and sat down
on the upturned skidoo to talk with my friend. We looked at
each other, staring into each other's eyes in the deepening
gloom of that crevasse. Ron was calm and totally accepting.
There was no pain now.

"Look," I said, "I'm going back up to rig up the winch and
I'll get this fucking thing off you and get you out, okay?"
And I leaned over the back of the skidoo and attached the
remaining rope to the tow hitch and the other end into the
back of my harness. I would use this longer rope to pull the
skidoo out of its position on my friend.

He smiled. He actually smiled and simply said, "If I'm
going to die, this is a great place to do it in. Life's been fan-
tastic and we've had some great times." He laughed and actually put his right hand over his mouth in the way he usually
did when he laughed and he screwed up his eyes again as he
usually did.

Probably, and I like to think so, some of the times we spent
together with the others of our tight-knit community on the
ship, had passed fleetingly across his thoughts. We had spent
all those months together on the journey down through the
Atlantic, and we had all been friends. Pulled closer together
in the spirit of our adventures, Christ we'd done enough to
raise a laugh.

We sat for a while not saying anything. We both knew it
was hopeless.

I stood up, knowing that I had now lost two very dear
friends and there was a third on the surface. I wasn't going

to let Nick die; I had to get back to him. But tearing myself away in these circumstances made me feel like a criminal.

As I turned to face the long climb out of the crevasse, Ron called me for the last time.

"I'm pleased it was you with me when I died. I know you've done your best. Thanks for being with me," he said. We shook hands, and I noticed that the gaping wound on his head was not bleeding anymore. I turned from him and stepped into my foot sling.

After I did a few hellish minutes of climbing, only twenty feet or so above him, trapped and crumpled by that machine, Ron, my friend died. I would dearly love to say that Ron died peacefully in that crevasse, but he did not. He was fighting when he died; he was fighting to the very end.

I hope to God that he did not die believing that I could get him out of that crevasse. I hoped to God that he forgave me for leaving him on his own in that terrible place. I hope to God that he has forgiven me now after all these years. And I hope to God that Joe and Ron found peace in that crevasse beside an unnamed mountain on that shambles of a glacier in Antarctica.

24

Days Out

During the winter I traveled with a number of people, normally confined to the base, on their journeys onto the snowfields and glaciers. Although officially called "training" runs, they were widely known as "jollies" and all the base personnel were given the opportunity of traveling, at least, across the most well used route to the disused base at the farthest side of the island. Only a days travel away, it was nonetheless still a serious undertaking.

My instructions before my first trip were simply, "Go up the ramp across the airfield, on to the glacier, turn left at the top. Head across the snow field until you get to the window on your left and then veer right on course two hundred degrees until you arrive at the slope overlooking the deserted base." Seemed simple enough, and it certainly looked simple enough on the map, but the instruction "the window on your left" intrigued me as it had done countless people before me.

I completed the trip many times during the year with various traveling partners and the route proved to be exciting, and although it was not a great distance, it still took a full day of hard traveling in good conditions. Experience of the drawbacks and also the luxuries of traveling on motorized transport were quickly gained, and my responsibility was the preparation of vehicles and equipment and the preparation of the route. The skidoos were loaded on the flat body behind the driver's seat with the survival gear and usually the bag containing sleeping bag and mattress. In addition, I always carried in this section a spare tent and survival sack with some food and spare fuel, ice axes, snowshoes, crevasse probes, and a shovel. On top of all this went my camera case and usually a flask or two of hot liquid.

Behind on the sledge were the main tent, boxes of supplies, and the main bulk of the fuel needed for the vehicles. The sledges constructed of willow were perfect for travel in these conditions, as they would flex and bend on the uneven ground.

On some days the traveling would be laborious and slow, as the front man would constantly have to dismount and probe the area in front for crevasses and to secure a safe route. However, time was not of the essence and the longer spent in the open was a bonus.

The run down to the deserted base across the snowfield was open and largely safe. On this terrain the fissures in the ice were easily seen and never very wide, allowing easy travel over their open mouths. The rock hard ice lips of the crevasses were rounded and worn by the winds and well spaced; with the constantly low temperatures of the winter this route provided superb training opportunities. With the mountain range to the left and the wind direction predominantly from the sea on our right, the route was easily navigated with an obvious path in clear weather.

High up on the escarpment of the rocky crag face of the lower mountain there was a huge, almost square hole through which the sides of the mountains behind could be seen. This was the famous "window," and all it lacked was a pair of drapes to prove what it was. It was such an obvious landmark, perfectly positioned to take sightings from, and a godsend after a hard day in difficult conditions.

The deserted base was approached down a long iced ramp and comfort was immediately found when the door was opened. Sets of bunks were set around a stove, and bookshelves contained enough reading matter to fill the hours of storm-lashed immobility. Small yachts had regularly visited this place over the years and their crews often over-wintered here, leaving supplies of French novels, no doubt in return for the English ones they liberated.

Wildlife teamed around the area with seals wallowing un-afraid, now that the humans who had been almost perma-nently based there had moved, taking with them the need to secure seal meat for the dogs. Penguins abounded, skuas, sheathbills, and various other birds flooded the area, and on a few occasions before the sea ice settled and formed we could see the blowing of passing whales close to the rocky coastline.

Wandering along the shoreline in the cooling temperatures of the late summer, many icebergs had begun the drift shore-wards for their winter of inactivity. The whole area took on a totally different perspective from that which I had first seen of the base in the mid summer.

There was an area where a yacht had been pulled ashore by long forgotten modern explorers some years before. This refuge on the hard ground of the shore had probably been used to stop damage from the encroaching sea ice during the winter period. A fence surrounding the area had been erected to prevent the seals lolling and rolling against the precariously balanced boat.

Days at this place were spent in peace and quiet, reading, sleeping, wandering around, and eating copious amounts of the dehydrated food that had been left behind in the evacua-tion. It was as though we sought this place to go into retreat, to bask in its quietness, and to experience the beautiful sur-roundings.

An impromptu pajama party was held there one night as four of us madly celebrated the end of one particularly hard journey. Although we did not have pajamas, we wore our well-worn thermal underwear, drank our limited supply of secreted beer, and moved around with candles on saucers like a tribe of demented "Wee Willie Winkies." That night we gorged ourselves on superb food and tinned Woodbine and Capstan cigarettes from the deserted store. These de-lights had remained at the base for years and years following the closing of the facility, and the tins of luxuries provided a welcome for many unannounced visitors over the years.

As usual our visits were all too brief, and we would move on because of our largely self-imposed time schedules. Regular radio contact with the main base would keep Morse, the Base Leader, aware of our position and status, and his relaxed style was superbly tested for this environment and our situation. I cannot remember seeing Morse lose his carefully controlled temper once during those long months, even when faced with the many great problems he had to endure. He had been captivated by Antarctica from the moment he first saw its outlines from the ship on which he arrived, and as a mountaineer he could identify with the freedoms we so wanted.

The journey back across the snowfields and glaciers was always anti-climactic in the knowledge that soon we would once again be in amongst the crowd on base. To the noise of coming and going, the constant drum of the generator in the background, the weekly chores to keep the base clean and tidy, and the welcoming howling of the dogs, we would return to our civilization.

But as always on return we would be allowed a shower, a luxuriating shower which was usually only allowed once a week. Even though at these times the traveling party would have the arduous task of replacing their water with freshly cut snow blocks for the melt tank. The time spent with saw and shovel cutting and moving blocks of snow was worth the effort as the grime of the travel was, once again, washed from our bodies.

Traveling was a time of pure joy amidst the wild almost untouched wilderness of Antarctica.

25

Journey: Back to the Surface

When Ron died, I stopped my climb up the rope and looked down. His head was slumped forward and his left, damaged hand was resting on the rail at the side of the skidoo that had locked him into his lasting position. His body, still contorted, was finally at peace.

Although I knew he was dead, the doubt remained in my mind. My logical side was screaming at me that Ron must have been dead. My conscience screamed at me that he could not be. I should retreat down the rope and pull the skidoo off his inert body and haul him to the surface, but I turned to the crevasse wall and began again the long difficult process of getting myself out of the crevasse.

Climbing upwards on the thin rope and sweating in the severe cold with the intense effort and the shock, I began to feel the exhaustion get hold of me and my arms began to ache with the effort. Perspiration running down my face forced its way down my collar and was hurried on its way into my clothing by snow dropping from the lip of the crevasse where the hole had broken through.

The thought of climbing past the knot joining the two ropes frightened me as the process of unclipping my jumar and then placing it on the rope above the knot seemed to me in my tiredness as an impossible task.

Trying to regain some strength and taking a breather just below the barrier of the knot, I looked to my right and left down the expanse of the massive crevasse. It disappeared like a huge freeway into the distance, wide and twisting on its inexorable journey to God knows where. The colors changed as light filtered through from the glacier surface above, through bridges formed over the gap. Some areas of the bridge looked frighteningly thin, and I felt I could see the sky uninterrupted through the inches of snow. If only the bridge that collapsed had been that thin we could have seen it before crossing and the accident would have been avoided.

I could not take in the sheer size of this crevasse. How could the gap at the top have been bridged with snow over a ten feet wide chasm? Secondary bridges littered the depths of the hole but at no point was I aware of being able to see the bottom. For all I knew there was not a bottom and the hole continued into the furthest depths of the earth itself until it met with the white-hot core.

I managed without any difficulty to get over the joining knot, and moved smoothly upward, praying that this was the last obstacle to my safety. As I continued upwards I could feel shivers spreading over my body as the further drop in temperature brought on by the end of the day began to cause my sopping wet inner clothing to cool. The further I climbed, the colder I began to get, even though the dropping snow had now almost ceased from its constant downpour upon me. I needed to look downwards at my lower jumar to move it upwards, and every time, fresh cold air slid past my collar and knotted silk scarf to hit my wet inner clothing with what seemed a terrific force.

Seeking a break in the effort, I stopped to tighten my clothing and to try to plug up all the gaps to my rapidly cooling body. But the stop made it worse, and the constant sweat and water drenched clothing would not let me warm up. Then I looked up and saw it. My body slumped back in my harness as I realized with horror the mistake I had made; the rope had disappeared into the groove it had made in the snow and ice of the lip of the crevasse. I had committed the most basic mistake of almost criminal proportions in my haste to get down into the crevasse. And with this criminal mistake I knew the outcome could result in capital punishment for me.

Rope will cut into snow and even ice when pressure is placed upon it; the constant movement of the weighted rope will act like a saw. I knew that and was well versed in preventing the problem occurring. Yet I had dropped the rope into the hole and had not placed a jacket or a bag or anything under it at the lip of the crevasse. This was the rope's only point of contact with the crevasse. The rope, on which I was depending for my life, had cut into the snow and ice and had disappeared from sight as it had been squeezed and compressed into its hiding place.

The effort now needed to pull the rope from the groove it had cut while supporting my weight so I could continue upwards would be massive. I looked at the problem in disbelief and then down through my free hanging legs, deep into the abyss, and felt for the first time that I, too, would now die here in this crevasse. The crevasse, which had claimed the lives of two friends, had not finished in its sinister work and was aggressively searching for a further victim. That victim was going to be me. I had fallen into its trap, not only once, but twice.

I imagined in my increasingly troubled state of mind that the rope was going to snap. I was convinced it was going to break as I ran my damaged hands up and down the rope as far as I could stretch to check for abrasions. The brand new rope used for the first time only a short time before was now going to snap with me dangling from it. I couldn't move, I was fixed to the spot, frightened to cause any bounce on the rope by pulling upwards.

I had never in all my years of climbing seen a rope snap, I reasoned as I fought for control. I had seen them frayed and cut. I had rappelled on ropes that had seen better days. I had trusted rope slings in the Alps, which could have been in place for years. And I had never seen a rope snap. Now hanging from a brand new rope in perfect condition my mind was tricking me into submission.

If I stayed here any longer I knew I would die, and I had to get moving so I hauled myself upwards to a point where the rope disappeared into its groove of snow. Repeatedly I had to thrust my ungloved hands into the snow as I tried to pull the rope clear of its hiding place, and my watch bounced on my wrist; my prized possession, my Heuer Daytona watch with its expensive look and bright steel strap. The watch that I was repeatedly offered money for in places like Cairo and Amman and Kinshasa would remain with me when I died, hanging from this rope. Perhaps in years to come, when my body flowed down to the sea as the glacier moved, someone would find the watch still working, attached to my left wrist. Perhaps they would wonder who its owner had been and how and why he had died, and I realized I had stopped working on the rope.

My battle continued. The rope repeatedly failed to come free from its groove in the snow and ice, and I cried with frustration and fatigue as with depleting strength I poured all my effort into saving myself. I shouted at the top of my voice for help, only six feet from the top of the crevasse, as my progress was once again stopped. I cursed and swore and shivered uncontrollably and cut my freezing hands more and more as I fought for every inch on that rope that day.

After gaining six tortuous feet of height I reached up and touched the top of the lip. My hands were almost useless now, frozen and swollen, but I knew that I could reach up and clip one of my jumars into the rope above the lip. The weight of my body transferred to the rope above the obstruction would enable me to breeze up to the surface, and I would be free and safe.

Less than five minutes after touching safety I was on the surface lying in the snow and the now deepening cold, face down, trying to warm my hands in the opened folds of my jacket, listening to the repeated pleas of Nick on the radio. I needed to get under cover. I needed to check on Nick and I needed to attach the rope to the winch.

I attached the small jockey winch to the skidoo on the surface. Then I pulled the rope tied directly onto the rear end of the skidoo that was jammed in the crevasse through the winch before engaging the mechanism. Two dead men snow brakes, flat metal plates designed to resist movement through snow, were fixed in place behind the skidoo the winch was attached to. I placed the dead-men snow brakes as deeply as I could, and I maneuvered the sledge into place on its side behind the skidoo as extra resistance. The lines between the skidoo and sledge and belay points were kept taught, and I began to pull back on the handle of the winch. Quickly the rope went tight as it took the strain, and just as quickly the complete belay system, skidoo, dead men, and sledge were pulled forward toward the crevasse. The weight of the skidoo lodged deep in the bowels of the crevasse had formed an immovable weight.

The only fail-safe system I had for crevasse rescue had failed. In my anger in the aftermath of the accident I had tried to make the system work, and at a time when there was absolutely no need to do so. And the futility of it all hit me. What was the point? Ron was dead, Joe was dead, and I was trying to pull the skidoo out of its jammed position with an inadequate winch anchored to a skidoo. I turned away from the crevasse to check on the only other survivor.

26
Mountain Rescue

I was already riveted to the program being shown on television that was minutely investigating the work of a mountain rescue team. My eyes and ears focused even more intently when the story of a difficult crevasse rescue unfolded before me.

As the furious activity of the mountain rescue team surged into even higher gear, it became apparent that they collectively considered they were facing a complicated, although perhaps not at this stage, losing battle. A young, fit, and experienced climber had slipped on the steep slopes of a glacier. He had gone into an uncontrollable slide for a short distance before breaking through the bridge of a hidden crevasse and falling into the gaping chasm. By crashing into and through a secondary bridge some thirty to forty feet below the glacier surface, he had slightly slowed his rapid descent before coming to a complete stop at a depth of some fifty feet. He was lodged on a stable platform of snow and ice, face down and with the remains of a further bridge creating a roof, of sorts, over his inert body.

His climbing partner, unable to raise a response from his friend and unable to effect the rescue himself, set off down the glacier in search of help. Within an hour or so the mountain rescue had dispatched two of their team by helicopter and placed them on the ground close to the site of the accident. After a careful traverse they located both crevasse and victim, and one of the mountain rescue team immediately began trying to establish contact with the trapped climber as the other put safety ropes in place and secured the site against further disaster.

Within a short period of time, two highly trained crevasse rescue experts were in position on the surface, and two similarly trained experts were in the crevasse, one of them with the casualty, and the other halfway down the hole organizing the rescue lines. The struggle became a battle against time as hypothermia and shock sank their teeth deeper into the victim, who had been quickly cocooned in down and Gore-Tex to stabilize his body temperature.

There then followed a desperate effort to get the victim to the surface amidst the pulling of ropes, shouted commands from surface to crevasse interior, and sheer hard work that took well over an hour. Eventually, with a massive effort, the rescuers and the barely conscious casualty broke into sunlight as the surface was gained and the waiting helicopter evacuated the victim at break next speed to hospital. Dropping low and very fast over the glacier, the aircraft skimmed the surface as the pilot used his superb skill by piling on the speed of evacuation to assist in saving a life.

The exhaustion of the rescue team, left to clear up the accident area of piles of used gear, was obvious. One of the team retreated from the crowd to be alone. He wept with relief at the effort and as the result of the stress of this horrific episode. Unfortunately, the battle was lost and the climber died shortly afterwards in hospital, to the complete shock of the medical and rescue teams. The climber succumbed to shock, the cold, and internal injuries.

When they were informed of the young man's death, the sadness and remorse visibly ripped through the rescue team, shocking all of them into dumbfounded silence and tears. Their effort had been massive and they had all become emotionally involved in the attempt to save this young man.

I sat and stared in disbelief at the screen, barely taking in the remaining parts of the program. With helicopters, masses of equipment, and experts in close attendance within a reasonably short space of time, this young fit man had died. And then another discovery hit me. I had only climbing ropes and a jockey winch attached to a skidoo on the surface as my rescue tools. How could I, on my own, have managed to affect a rescue in the circumstances we encountered on the glacier that day?

Over the years since the accident, I had constantly berated myself at my inability to extricate the two victims. It had never once occurred to me that I could or should have done anything less than get the two victims to the surface, even though I had been fighting a losing battle in desperate circumstances. Mountain rescue team members, well provisioned and well rested and taken by helicopter to the scene without any real emotional ties to the victim at the scene of the accident had failed to save a life.

Countless books on the subject of mountain rescue state in their pages that when someone falls into a crevasse, they can be pulled out using a variety of techniques. Unfortunately the circumstances of the accident I was involved in on the glacier were not as straightforward, and not once throughout the years since had I really considered this. I considered I was at fault and tried to live with my guilt. I hung the blame around my neck like a pendant and even though the facts were staring at me, I was destroying my life. The reality was that traveling on a skidoo in crevassed terrain is treading a very fine line between being safe and not being safe. Ron had been un-roped that day on his skidoo, but he fell with the machine. Joe was roped to the machine but he also died. In the aftermath of the accident I could find no logic for traveling with that type of transport.

The odds had been stacked against our party in that devastating accident from the beginning, and I felt that the Shambles would never have given me back my two companions no matter what I had tried.

27

Journey: Radio Contact

Nick continued to trawl through the radio frequencies appealing for help as I put the tent up around him. He was dejected and cold in the deepening gloom of the day, and with sapped morale, fear, and shock, I knew he was in danger. The warmth of the tent would help as I tried to make sense of our situation and the mess of gear lying strewn around the site.

After securing the site, I unclipped from my safety rope and lent down to enter the tent that was now quiet. As I stooped down to crawl through the tunnel entrance, a ball of flame and metal flew through the opening and hit me full force on my head. Ducking and rolling I checked for fire on my clothing and then looked up to see Nick's head and shoulders peering out of the tent. We laughed out loud, releasing our tension in the adversity, as he doused the flames on the primus stove in the snow.

During a break in his attempt to make radio contact, he had begun the process of making a hot drink for us, but discovered we had finally run out of paraffin fuel. So he had used skidoo fuel, a lethal mixture of gasoline and oil, to refill the stove. In his haste to get the stove burning, Nick had opened the control valve slightly too soon and the un-primed mixture had escaped from the jet in a very fine and powerful spray which immediately burst into a bright orange ball of fire. As he was sitting inside the tent and the stove was blocking his exit, he picked the burning mass up with his bare hands and threw it through the open tunnel of the entrance before the flames could ignite the tent. Unbeknown to him I had just begun my entrance. Aware of the state I was now in, Nick reprimed the stove and brewed in record time, not speaking to me or encouraging me to talk until we both held hot drinks.

As the steam from the cups clenched between our hands escaped into the roof of the tent I looked straight at him.

"They're both dead," I said, and took another drink from my mug before telling him of the horror down in the depths of the crevasse. Nick held my forearm, leaning hard down on it with both of his hands when I told him of my brief conversation with Ron, and he turned to the radio once again and drummed it into life.

I felt empty and exhausted. My chest felt as though it had a huge hole in it. I sat and watched my hands thaw as I held them over the glowing flames of the stove. I became aware now of the many cuts I had collected from banging my un-gloved hands against the rock hard ice and snow of the cre-vasse, and both of my hands stung now as the blood flowed back into them. Within an hour, the pain was so bad I had wanted to cut them off.

Nick had worked the radio nearly all this time with the conviction and determination of a classical musician playing an instrument during a long practice session. He had tried everything he could think of to raise a response from some-one, somewhere out there, and in his concentration he had not heard my shouts when I was just below the lip of the crevasse, only a few short feet away from him.

His frustration and tiredness were readily apparent as we both cursed the base for not keeping a radio watch, even though we had never expected them to do so. We could only wait in frustrated grief, as it was now obvious that the regu-lar schedule would be our first possible contact. By this time it was approaching 7:30 p.m. and the time of our regular radio schedule. A crackle from the set broke through the silence, and we both heard the tones of our popular radio operator coming to life. Nick took the handset. With a calm and controlled voice, and without going through the process of confirming signal strength, Nick began. "This is Nick, Sledge Golf. Please get Morse to the radio and empty the radio shack of everyone else, over." Without hesitation the radio operator replied. "Copy that, Nick, wait one minute and stay on line, over." A few minutes passed and the radio crackled into life again.

"Sledge Golf, this is Morse, go ahead, over."

I took the handset from Nick as we had decided and I began to blurt out my message.

"Morse this is Steve. At around 4:00 p.m. today, Joe Watson and Ron George were both killed after falling into a crevasse. Both remaining members of the party are safe and well. Our exact position is not known at this time but I can confirm that we are at the top of the glacier and off course, but within a short distance of the correct route, over."

There was a slight, almost imperceptible pause at the other end as Morse, no doubt, tried to take in the enormity of what I had just passed over the airwaves to him. He was the base leader and the inconceivable had, within the last few hours, happened to members of his team.

"Copy that, Steve. Can you confirm that both you and Nick are okay, over."

"I can confirm that both of us are okay. The tent is up and we are going to get some food into us as soon as possible, over."

"Copy that. Please keep this line open at all times. I will be back to you within a few minutes, out."

And with that, the man was gone to inform, by code, headquarters in England and to organize and inform, face to face, the "by now" assembled guys on the base.

The man was brilliant during those few minutes of dialogue on the radio in a way that is hard to describe. Morse said the right thing at the right time and showed stoicism I had hitherto never experienced. But more than just remaining calm and doing his job, in those few short sentences of dialogue he calmed me and reminded me, with his calmness, that I had a responsibility to Nick to get him out alive and in as good a shape as I could. He filled my body and my soul through his true leadership qualities with a newly found confidence. I visualized him talking to me, and I could see him talking me through this catastrophe. I realized that my role was not yet finished. I had to find the strength to continue, at least for the time being.

Several years later I tracked Morse down. I telephoned him but for some reason I couldn't bring myself to tell him how much his words in control had helped me to get through. We did not touch on the accident during that conversation.

As we waited for the radio to come back into life I began to give Nick more details of what had happened and what we had to do. The shock was palpable now in both of us, but we had to survive. And I think both of us prayed at the time for good weather.

Morse's voice came over the radio again. "Okay got your situation. We are organizing a party to come and collect you. They will be leaving first light tomorrow morning. Can you confirm that all is okay, over." And on it went into the now ever darkening night.

The doctor came on line to check our condition. The radio operator came on line to check the radio would not fail at this time, and the line was held open as the dialogue went back and forth, making arrangements for the evacuation and letting us know who was detailed to journey out to meet us.

Nick had arranged the sleeping mats and sleeping bags and he crawled into his down cocoon for the first time since the accident to try to get some warmth back into his tortured body. The primus, with its gasoline and oil mix fuel, roared its now friendly roar and the temperature of the interior of the tent was soon rising in sharp contrast to the deathly cold outside.

We ate a meal and sat and talked quietly. I left the tent and rigged up a new radio aerial, stretching as far as I could stretch the antenna in a V-shape. My steps took me out into the gloom, downhill with my head-torch burning brightly in the cold crisp air of the Antarctic winter evening. It was only later that I realized that I had not attached myself to a safety rope while carrying out this operation; I just couldn't believe my stupidity.

Our dialogue with the base over the radio had been arranged to continue throughout the night with a relay of people at the other end. But by midnight the conversation was drying up and Nick had fallen into a totally cocooned sleep, with no part of his body showing above the opening of his massive red sleeping bag.

The radio communication was broken off, and I lay back on my sheepskin covered air mattress before getting myself into my sleeping bag, ready to enclose myself in the folds of nylon and down and hopefully sleep. The lamp and the stove were now extinguished and the temperature was rapidly trying to catch up with the temperature outside. My mind wandered and I could hear Ron shouting to me from the confines of the crevasse. He was calling to me clearly, he was okay but he needed help to get out. As I jerked forward and moved toward the tent entrance the calling abruptly stopped. I waited and lay down once more as the calls started up again, and this time the voice of Joe joined in, laughing at the predicament, a laugh I had so often heard in the past.

I knew it was the wind and I knew it was my imagination, but the doubt was beginning to infest me and tease me, and it was within only a few minutes that I decided I had to go back down into that crevasse. I had to get back to the scene and confirm that I had done as much as I could have. So, I quietly covered myself in my protective clothing that had festooned the inside of the tent. The normally quite difficult process of getting dressed in the close confines of a tent this time went by in a flurry of quiet poignancy, and in a short time I was ready to leave. My excuse that I needed to check the tent was at the ready in case Nick stirred at my noise and woke. But he didn't move in his tight cocoon as he fought his battle for warmth and sleep and I slipped, as noiselessly as possible, into the open night.

My head torch cut through the darkness as I immediately felt the intense cold of the clear, still night, and I tied into a safety rope as I began the process of sorting my equipment. After tying into the rappel rope and attaching two sets of jumars to my harness, I edged towards the lip of that bastard of a crevasse once again. Putting a canvas bag under the rope at the lip, I also clipped a further spare jumar into the rope above the lip with two long blue tape slings attached, so the extension dropped well down into the blackness of the hole, creating another foot loop to aid my escape.

My clothing squeaked against the hardening snow of the lip as I lowered myself downwards, and only seconds later I was at the knot joining the two ropes. The eeriness played havoc with my mind and I almost turned tail and fled, as the shock and the guilt coursed through my veins. Bright light from my headtorch bounced off the sidewalls of the crevasse in blues and greens and white, and the vapor from my warm breath was caught in the beam of my torch as I glanced from side to side. The same light exposed the horrendous scene. Hanging from the rope I slowed and came to a stop with my feet only inches above the stricken skidoo.

Ron was in exactly the same position, as I always really knew he would be. His body unyielding as I reached over with my ungloved left hand and felt beneath his collar for a sign of life. He was cold now and his well muscled body seemed incredibly fragile in the freezing, still, silent air, but he looked exactly as I had last seen him with his head bowed forward, although his eyes were now closed. There was a thin layer of ice on his eyebrows and the stubble on his chin.

I told him I was sorry that I could not get him from beneath the heap of metal pinning him in this place. And I told him how desolate I felt at having had to leave him alone to die. With one last glance in the direction of Joe, who I couldn't see as his body was obscured by machinery, I turned back to face the wall of the crevasse and I began my long depressing climb back to the surface.

They would be comfortable in their eternity down here.

My aching arms felt they were being wrenched from their sockets as I climbed the rope.

Tiredness was fighting to cover me now like a blanket as I reached the loop in the sling attached to the topmost jumar just below the lip of the crevasse. That sling saved me minutes of struggle, and soon I was once again surrounded by night. The stars were blinking crazily in the sky for as far as I could see, and I felt the intense cold of the snow on my legs as I knelt at the lip of the crevasse. But now the intense cold of the windless night against my sweat stained face threatened me. I coughed involuntarily as I heaved in lung full after lung full of freezing air. My body could fight the trauma now and I felt inside me a fierce determination to make certain we would get out of this place in one piece.

The tying and untying of lines was carried out in automatic gear as I sorted through the various ropes weaving their scribble patterns on the surface. Like a bell sounding, the clicking of my karabiners shattered the perfect peace of the night, so I stopped my work only to hear that the noise of my boots on the hard snow was loud enough to wake the dead. Hanging in the sky, the large moon was casting a faint, doleful light on our camp and our cemetery as the stillness stretched forever, over the glacier and up to the tops of the mountains and beyond. Surely, there was no one left in the world now to interrupt the peace.

Throughout those hours, I was aware of the massive presence of the mountains closest to us and felt that perhaps we had strayed so far off course the crevasse which had claimed my friends was the bergschrund at the base of the mountain, the point at which the rock of the mountain joins the ice of the glacier and creates a gap. It was academic; it was a hole in the ice all the same. A hole that had been big enough to swallow equipment, vehicles, and friends. But we were too close to the mountain, that fact was proven. The ground sloped to such an extent that I should have realized long before I did that we were off course and in danger. A tragic mistake, the consequences of which were now all too apparent.

The cold, even in the still air, was forcing its way into my clothing, hurried by the sweat on my body, and I finally retreated into the tent to try to catch some sleep before the sun rose again. I don't know what time it was when I managed to get into my sleeping bag, but it seemed like only a few minutes before the radio, lying by my head in its position of prominence and importance, burst into life again.

28

Discoveries

Many years later, during a flashback, my thoughts turned abruptly to Nick. I thought of the scene he had witnessed, and the amazing strength he had to find in order to stay alive in the immediate aftermath of the accident. Of small build, Nick was a very strong willed person who had lashed me, and others, with his strong accent when discussing topics close to his heart. He had traveled on occasions the previous winter, and although the terrain was not totally unfamiliar to him and although he was aware of the potential dangers, he had little experience of lie ups and was simply not prepared for accidents.

As he traveled, following in the tracks of Ron's skidoo that day, Nick watched with absolute incredulity as within seconds the accident was played out. He stared ahead as Joe, the passenger on the rear of the skidoo, looked straight at him with raised goggles and his mouth wide open in complete astonishment as they began to fall through the bridge, still holding tightly to the guard rails and in his half kneeling half sitting position. In an instant, both Joe and Ron were gone, and the fully loaded sledge they were pulling rushed forward noiselessly and followed them down into the crevasse with the speed of the falling skidoo.

With a superhuman effort, Nick, shocked in the trauma and the cold, recovered sufficiently and he poured his depleted energies into trying desperately to raise help on our radio. He tried to contact almost every station he knew of while he transmitted an SOS and scanned the frequencies. Eventually having contacted our base, we faced each other inside the protection of the tent on the still uncovered bare snow floor, and he recounted to me what he had seen as the skidoo fell through the bridge of the crevasse.

It had been as I had initially thought, having seen intimately the size of the hole, and the beginnings of tracks on the other side of the bridge. The skidoo, with its two unsuspecting passengers, had almost cleared the danger zone but had then been dragged backwards into the hole as the bridge fractured directly under the rear of the vehicle. The crevasse bridge had collapsed as a direct result, in Nick's opinion, of the weight of the two people aboard.

Within minutes of the event taking place, Nick was in almost total shock. As the weather deteriorated it made it more difficult for us to survive, and the cold began to cut through our clothing. Combined with the colossal impact of the shock, my now single remaining partner undoubtedly would have become almost hypothermic in a very short space of time.

Many accidents that have happened to mountaineers and travelers in cold places have ended in catastrophe due to the onset of hypothermia. People in a seemingly safe position following involvement in an accident but without apparent injury themselves have succumbed very quickly to the drop in the temperature of their body core. I had witnessed the effects of shock when involved in a rescue some years before in the Cornish summer. The weather had been bright and sunny, and there was a very slight breeze coming in off the sea as I climbed a reasonably straightforward route, wearing only shorts and climbing shoes. On my way up the route, I approached within twenty feet or so of a couple of climbers fully equipped and roped, attempting a route running parallel to that which I was climbing. Suddenly, the lead climber fell from the vertical cliff above a long and seemingly well protected corner, but the force of his fall stripped out several pieces of protection he had placed in the crack. He plummeted fifty feet, landing on his back at the feet of his partner on a ledge.

Making my way to the scene, I found the partner of the victim had immediately gone into shock. With his hands crossed in front of his belay device and in classic belay stance, he remained rooted to the spot. A passing climber at the bottom of the cliff had also witnessed the accident and immediately sent someone for help as I turned my attention to making the site safe and lowering the uninjured member of the party to the bottom of the climb.

The victim, with broken limbs, was taken off the cliff by the rescue team as I rappelled down and rejoined the still staring, unmoving, and unspeaking partner of the unlucky climber. He was shivering almost uncontrollably, even though he was now wrapped in warm jackets, and he accompanied the victim in the ambulance on its journey to the hospital, where he remained for twenty-four hours. Even though he was dressed immediately in warm clothing, he remained freezing cold for a long while afterwards as people continued to sunbathe on the beach close by.

At a temperature fast approaching minus thirty degrees centigrade, the situation at the accident site in Antarctica was becoming more critical by the second. Although the accident filled my conscience and the horror of our predicament burned into my brain, I was aware of the need to protect the survivors, myself included, from further harm. I was aware Nick was suffering dreadfully, and I am certain he survived due to his remarkable courage and fortitude and his desire to hang on to life when the cards were being increasingly stacked against us. Due to his enormous perseverance, he managed to raise the base after more than three hours of radio transmitting. At this point he traveled into his personal and most dangerous time; he relaxed with final relief and due to massive fatigue, the shock, and the cold, his body temperature must have plummeted to the point where he could so easily have become the third victim. But he continued to fight and immersed himself in his sleeping bag, gradually warming his body through as he slept in the aftermath and we waited for the rescue party to reach us some hours later.

29

Journey: Rescue

Scanning the escape route through binoculars as it disappeared into the distance down the glacier, I knew we were entering yet another dangerous time. Nick and I were both gripped with massive fatigue. From my position standing on the seat of the skidoo I was praying that I would catch site of the rescue party, but they were hours away as the morning was just upon us, and my entreaties for their super fast travel were impossible for them to answer.

The cold was biting at us and was having a more profound effect on our depleted morale as Nick struggled once again to make a hot drink. His disappointment was apparent when I told him there was still no sight of relief.

Discarded equipment left where it had been dropped over the previous hours of fear and panic meant the inside of the tent was an absolute mess. The skidoo fuel we were burning in the stove created a black smoke during the priming operation, which had dirtied us on our hands and faces, and the smell permeated the atmosphere of the inner tent. But the radio stacked on a wooden box took center stage as it crackled into life once more.

"Sledge Golf, come in, over." I grabbed the handset. The party had left with two aboard and was heading towards us in good weather conditions, and I chatted with the Base Leader and gave my weather report in well practiced style, confirming that the visibility was excellent with no cloud cover. Once again I repeated our location as the party coming towards us would radio back to base every hour or so.

I gave the handset to my companion and left the confines of the tent ostensibly to try to create some order outside, but knowing that I would once again glance anxiously down the sweeping tumbling glacier to will the party to travel faster on their journey and into sight. My eyes kept looking over to the ropes attached to the skidoo and disappearing down into the depths of the crevasse. Even though I was attached to a safety rope, I did not go close to the accident site again, and I was to leave the task of releasing the knots and dropping the ropes down the hole to someone else. Someone who had a clear guilt free conscience and someone who could summon the courage to cut off the final reminder that friends were one hundred feet below in their newly found deaths.

Midday came and went with radio calls informing us of the progress of the rescue party. They had reached the bottom of the glacier but had hit some difficult ground and had to spend time probing the area to find a safe passage. The weather remained at its brilliant best as I prayed that it would remain so. After the last few weeks I would not have been surprised for a quick change, but today of all days it looked positive, and the good visibility and lack of wind continued.

Nick's morale was very low now and although, following the accident, he had rallied and performed superbly, he now only wanted safety and to be out of this place. He drank more hot drinks and tried to laugh at arguments and comments made during the previous days of confinement. But it was too hard; the shock and the grief, and in my case the guilt, were the dominating factors. We were barely alive, although I didn't recognize it at the time, as the supreme effort we had made to save ourselves had simply caught up with us.

The radio crackled through into an open line once again. Our rescue party, now talking directly to us, were within striking distance but they could not see us. "You must be able to see us," I said beginning to panic. "Look up to the left of where you're standing," I said but they came back with a negative.

My companion grabbed my arm. "Put a flare up!" he said, and in my confusion I replied that I didn't know if we had any. "What do you mean?" he said. "You are responsible for the survival gear. You must have packed them!" Again panic cursed through me immediately. Had I forgotten to pack the flares? Of course I had packed them. We had lots of flares, but in my confused state I simply did not know what he was talking about. I knew exactly where the flares were.

"Tell them to look up to mountains on the left," I said as I ripped open the bag and set off the first smoke flare. The response was a no sighting, so I immediately let off a "Very" light and watched as it shot up and exploded and traced its red arc across the sky almost directly above our heads.

I dived to the entrance of the tent just in time to hear, "Got you." Then a pause. "Christ you're high up." My world collapsed like the bridge of the crevasse that had collapsed under our companions.

The party approached us carefully, and in my impatience, I remember thinking too carefully and too slowly, as the front man in the party dismounted time and again to probe the ground. It took an age for them to cover the last remaining four hundred yards to our mess of a hurriedly erected campsite. Then, with a roar and an immediate cutting of engines they were upon us. I slumped onto a sledge and waited for them to disentangle their skidoo safety ropes and set up new safety lines in order for them complete the few remaining yards on foot as they approached me side on.

During the few minutes it took for them to cover the distance between us I considered deeply my immediate future and what would happen when the first words between rescuers and rescued were uttered. In my guilt I imagined that I would not even see the first blow as I was to be punished for putting ourselves in this position and endangering the lives of those who had come to help. I knew I would not have time to blurt out my defense of crazy weather conditions, damaged machinery, and depleted stores, as well as the tiredness, which now gripped me to the depths of my soul. I knew I would not have the time to express my sadness at losing two members of the sledging team and I knew that the pleadings of our maniacal efforts to keep the two remaining of us alive would fall on deaf ears.

But I had to prepare myself, so what. Surely nothing worse than that which had happened could be showered upon me in this world or the next. I had been to hell and back during the last thirty hours and I had little energy left to fight.

Wilfred put his hand on my shoulder with a brief word, while the second of our rescuers fumbled with his uncovered hands in his clothing before thrusting a lighted cigarette into my mouth. I took my first long drawn out drag and filled my lungs with the intoxicating smoke. Looking straight up into his broadly smiling bearded face, I waited resignedly for the pain of the blow. Instead he, too, touched my shoulder and said, "I knew you wouldn't stop smoking."

There was an urgency now. Everyone wanted to get away from the scene as quickly as possible as the travel up the glacier, I was to learn later that day, had been pretty hazardous.

I threw our survival bags onto the sledges and roped them down and we picked up some personal gear and were ready to go. The tent was left in position with all the remaining equipment scattered around the inside and littering the landscape outside. The race was on to get away.

My eyes followed the line of the ground from the mayhem of the glacier up to the beautiful ridges and peaks. The mountains, demonstrating their awesome power, seemed to look away from me, casting a veil of innocence over the scene and denying any responsibility for the tragedy. Perhaps, I thought, they were merely contemplating more important things than those that had befallen our beleaguered traveling party. I knew then that my greatest problem was going to be understanding, how such an outstandingly beautiful place could be the cause of such misery and deluge.

I forced myself to take one last look at the hole in the bridge of the crevasse as Wilfred, the lead man in the rescue party, reached forward and separated the ropes from the skidoo that were attached to our friends. As he dropped the ropes downwards, this former military man put his hand to his head in salute. In that movement we had lost them forever.

30

Journey: The Last Lap

There was nothing left to do now except retrace the route back down the glacier. We would head out over a few miles of snow field, approach the wide open expanse of the snow field which housed the aircraft runway, achieve the ridge on our right side, follow it for the two or three miles of its length over well established track, and we would arrive at the top of the long slope which led down to the base. The whole process would last no more than a few hard hours.

As we set off, I viewed the scene from my seat at the rear of the rear sledge. I was the last person in the train with the nothingness behind. The weather was beautiful, beautiful and brilliant with a visibility that stretched for miles in all directions.

With boring regularity the skidoo train would stop, engines would be cut, and the lead man would dismount, tie on with a long safety line, and move forward to probe the ground, which was soon to bear the weight of our passage. He was being super cautious now and rightly explored any of the depressions he caught site of in the path he was forging. I was no longer in charge, and it was no longer my responsibility to interfere with the route finding.

The rescue party, on the way out, had left markers in place, and the route did not deviate from the path they marked. The crevassed glacier lived up to its name as it poured down from the heights of its origin in a broken and tumbling mass of irregular ice, without any sign of the moraine of jagged and fallen stones that litter the surface of glaciers in other parts of the world. Through the miles stretching out before us, I could identify the gap down and to our right where the skidoo train would take a sharp right turn and immediately climb the steep, short incline to regain the plateau.

After the length of the intervening years, my memory of this short stage of the final part of the journey continues to remain in a haziness that I doubt will ever be fully cleared. But I do remember feeling secure on the sledge and safe in the knowledge that our saviors were on the top of their very extensive and skilled game. Surely there was no need to worry about anything now as the extraordinary shambles of a glacier disappeared to my left and then behind me as we climbed up the short ramp to gain the plateau.

During a break, sipping hot tea, I remember accepting gratefully yet another cigarette. It was the last one I enjoyed for a very long time as the inhaling of the smoke immediately caused me a nausea that was to last for many hours. The scenery lost its interest for me as I began to, self pityingly, concentrate on the feeling of illness that was now surging through my exhausted body. It was all I could do to stay aboard the sledge, and I had to hold on tightly until the snow eventually began to flatten out as the going became smoother and the speed was increased.

Familiar places, familiar landmarks began to appear as we traveled. The snowfield where, only a few short months before, Joe had experienced his first night under canvas through the hours of an Antarctic night passed beneath our tracks. Joe had not traveled away from base over the landscape until then. As base cook, he had arrived for handover from the cook who was due to return later that summer, and he was thrown in at the deep end. His predecessor was going off on his well earned post-winter holiday a few hundred miles south of the base for rest and relaxation before he would depart in the northbound ship at the end of the summer.

Much activity was taking place and there were a number of very hard working and regularly hungry people to attend to. Joe's days were totally filled, but even then, whenever I was on the base and not traveling, he appeared calm and relaxed and never in a hurry. He would wander around, occasionally ducking into the kitchen to check on something or other, before wandering again. The meals were always on time and the meals were always good. When visiting parties of foreign Antarctic workers arrived, Joe did not panic, and when the Royal Navy dropped by he was the same. He just got on with his job.

With the departure of the summer personnel, opportunities opened before him. He could leave the close confines of the base and leave those left behind to get on with it. And he eagerly anticipated his first challenging journey into the white unknown.

We left the base in superb weather on his first trip. Joe had practiced his self-rescue techniques and was comfortable and reasonably confident that this trip was a perfect excuse for him to sharpen his skills and to equip him with some experience of travel. I had decided where we were going to spend the first night, but as it was quite close to the base, within a mile or two of the airstrip, I made a long detour over relatively safe ground. I pointed out features as often as I could during our regular and long breaks, and my traveling partner appeared to be enjoying the freedom. From a very early stage I learned that he expressed his joy by shouting and laughing out loud.

We could have still traveled on for another two hours when I made our first camp in the wide, open snowfield, but it was still reasonably late and in the last throes of the Antarctic summer winter transition. The sun would go out as though a switch had been flicked, and I wanted both of us to enjoy the experience without taking on unneeded and additional pressures.

Lying back in the tent after our evening meal, we were chatting and looking through the opening of the tent. The perfectly visible tracks of vehicles threaded there way in a curved line in the sunlit snow out into the distance. From nowhere, we were hit for a period of about ten seconds by a tremendous katabatic wind; a wind that rolls from the mountains as hot air rises and collides with the cold air of the plateau before speeding out of control in a revolving ball of ferocity.

It was so powerful that I thought my plan of camping so close to the airstrip had been discovered. I thought one of the people working there, securing the airstrip hut for the winter, had seen us, crept up, and grabbed the outside of the tent, shaking it vigorously to frighten the occupants.

Joe looked across at me. "What the fuck was that?" he screamed as the wind disappeared just as quickly as it had hit us. He then dived through the entrance and looked around the tent; there was no one there. I tried to appear nonchalant about the phenomenon, but was just as shocked, and although I knew what had caused the tent to shake so violently it had been the most frantic katabatic wind I had experienced thus far.

The next few days were spent in brilliant weather until we got to the abandoned base at the very furthest extreme of the island. The slope, approaching the weather beaten buildings in the distance, was blown clean of its covering of snow, and as it was the edge of a glacier, the sheet rock hard ice underneath had been exposed. The only way to gain shelter for the night was to abandon the vehicles, put crampons on, and trudge the remaining distance down the slope to the safety of the buildings.

We made a couple of journeys by foot that evening, ferrying our equipment, and had a superb four or five days lounging around and taking in the magnificent sights and wildlife.

Close to the shoreline were several large, impressive icebergs. Some of them caught by the shallowness of the water seemed to fight for their freedom once again as they ground noisily against the rocky bottom of the bay, rising and falling almost imperceptibly in the light swell.

But one towering ice structure, fifty or sixty yards offshore, took on the grandeur of a small cathedral with towers, ramparts, and flying buttresses, creating its own edifice to nature. A sea mist surrounded its lower sections and caused an eeriness in the slowly ebbing waters of the dim calm evening. Fifty yards behind this cathedral and slightly to the right was an iceberg that only confirmed my feelings the other was indeed a cathedral. Floating freely there was a solitary mound of ice. Pointing upwards almost perfectly triangular in shape, and perhaps fifty feet high, it outlined the shape of praying hands with palms forced together. My imaginings perhaps, but the second iceberg added a solemnity to the occasion and a credibility to my belief in the iceberg cathedral.

Now, some short months later, I was the rearmost person on the rearmost sledge traveling across the very same snowfield on which Joe had first journeyed in Antarctica. Traveling across ground on which we had traveled together a number of times since then. Now he would not travel over it again.

The leading edge of the ridge came into sight as we were running freely and safely over the well-used snowfield that was the aircraft runway. Barrels marking out the landing strip for the airplanes remained in their places, awaiting their removal for winter. The small red hut that was the airfield operations center dominated our view. Two miles from home.

As the ridge powered upwards to our right, the snowfield airstrip was left behind us, and the familiar right to left slant of the terrain allowed the sledges to drift downhill very slightly. Within a short distance of nervous travel, there ahead in the near distance, the rock and ice covered buttress signalling the top of the ramp appeared.

Familiar buildings nestled in their permanent place below and to our left, surrounded by the snow-encrusted ground, scarred by the movement of people and vehicles and dogs. Excited dogs that had picked up the familiar sounds of vehicles approaching, probably from miles away as the noise drifted across the uninterrupted air, barked and yapped. Pulling at their traces, all the dogs were circling in their excitement, unaware that two of their friends, companions since mid-summer, would not be seeing them again.

Against the backdrop of pristine new snow, the flat rub-berized fuel storage tanks, out on the promontory, were ly-ing in position beneath the single raised flag on its white flagpole. The journey was almost over and home was within strolling distance.

The skidoo train stopped at the top of the six hundred meter slope leading down to the base as the remaining sun blazed in all its glory without a cloud in the sky to spoil its radiance. Islands eighty miles away in the south were clearly visible in their mirage in the bitterly cold, clear air. Sea ice, which had only been forming into a more stable mass when I last saw it, a few weeks before, was now firmly established in one continuous white sheet right across the bay to the other side some eight miles into the distance. I clearly re-member four or five seals lying on the sea ice by their es-cape holes, taking in the last rays of the sun, and I remember how the drivers of the skidoos, heads down, unsmiling and concentrating, went about their business of reorganizing the sledges.

Within thirty minutes the cold would be closed out and we could crack open the iced up zips and strip off our protective clothing layer by layer. The feeling of the ice thawing around our faces and rivulets of water dropping from our beards and long hair would tell us we were home and safe.

One skidoo in front, two sledges one after the other be-hind, and the rear skidoo acting as a brake at the rear of the train were all roped together. We two survivors reclaimed our positions on the sledges and the train started again on the very last stage of the rescue journey.

The going was slow and careful, with the rear skidoo braking hard and only being pulled forward by the power of the first and the momentum created by the steep ramp. In the distance, past the train, friends were now gathering outside the base, looking upwards with covered heads and gloved hands. They were friends who I had to face, with my guilt, for the first time since the catastrophic events of thirty-six hours before. They were friends who had lost friends.

As we drew closer, a softly spoken word or a gesture from the Base Leader sent the others as one back into the green building that served as our home. The Base Leader remained alone, standing in front of the building, watching our progress.

When the skidoo train finally stopped, no words were spoken. The two drivers closed down the engines, dismounted, and walked away with their own thoughts, while Nick walked disconsolately towards the door, hesitating only briefly to shake hands with Morse before moving on, head bowed. Safe.

I dismounted with more trepidation. Slowly I walked towards Morse, trying to read his eyes. I didn't have the time to assimilate as he strode purposefully and put his arms around me with a bear hug, and I wept like a child. I didn't have to tell him what we had been through. He knew. He was a giant of a man.

Shock must have been coursing around my body and the relief of seeing Nick on safe ground had forced my resolve to disappear. I knew the questions that I would have to answer could prove to be more than I could bear in my present state as Morse's grip was released and we trudged slowly, not speaking, towards the open door. Once through it, I came face to face with the generator mechanic who in time was to become my main source of strength; the man who pulled me through the initial days and weeks and months. Our eye contact was brief but I was able to see in him the manner in which the base had pulled together to mount the rescue.

From the optimism of looking forward to the usual formal Saturday evening meal, retained as a tribute to the times of the real polar explorers, the hell of the situation had been presented to these men. The joy of the Saturday evening relaxation had been kicked out of them during the short time it had taken me to transmit the news of the tragedy. This day had been spent worrying if any of us, including the rescue party, were ever going to reappear at the top of the ramp again. The horror of accident and death of this magnitude was a situation new to all of us, and any reaction to such a situation was totally unknown territory. All the members at the base were exhausted after having spent the night servicing the rescue skidoos and preparing the sledges for travel in the emergency; few had had little or any sleep the night before.

The first inquest had to begin. The Base Leader, the doctor, Nick, and I congregated in a blur of wet clothing and steaming bowed heads in the room that acted as the surgery.

My words could not leave my mouth. I was caught in the aftershock of grief, horror, and self-pity, and the big orange toes of my heavily insulated traveling boots became the sole focus of my attention. The minutes of total silence seemed to drag by as the water dripping from my hair made a pool as it splashed to the floor, and I avoided the gaze of the others and tried desperately to regain some composure. I was the only one who knew the truth, the real story, and now I was incapable of telling it.

The doctor stood in front of me. He captured my attention, and deftly put a small pill into my hands. Mechanically I put the pill into my mouth, and within seconds I could feel calm surging through my veins as the drug hit my system with almost immediate effect, and I regained the strength needed to begin the horror all over again.

Dispassionately, I recounted the story. And with every word I grew more worried, as the drug took greater hold of me and so flooded me with its relaxation that the listeners would think I didn't care. Not a hint of emotion slowed me down and I felt as though I needed to keep dropping my head to try to reassure them that in fact I did care; that I had lost two very close friends and that only by the grace of God, two of us had survived one of the worst accidents in the history of this place.

The questions were asked when I finally stopped my uninterrupted monologue. The doctor remonstrated that Ron would have been unable to survive the injuries that I had described in the most lurid detail. That in their shock, both

Ron and Joe would have died in peace. Never once were my decisions questioned and more questions were asked and answered in that room as we stood in shock and grief. A minute, an hour, or two hours later, I have no recollection of how long we talked, I turned and left the room and aimed my body toward the showers. As the attempts at sympathy and understanding followed me through the door, the words made no impression on my back.

That evening is lost in a blur as the whole base met and watched a film, usually only allowed once a week. Nick dealt with his immediate shock and grief in his own way and bounced against every wall as he staggered down the stairs outrageously drunk later on in the evening. I remember looking at him and grinning sadly as I followed his ungainly progress. We, the two survivors, had caught each other's eyes once or twice that evening, but not a word had passed between us since leaving the sledges after our final journey.

We rarely spoke to each other during the remaining months of our trapped existence. Letters were exchanged between us only once a few years later, but no mention was ever made between us of the trip that ended so disastrously. We were left to get over it and get on with life, and I suppose we tried, each in our own way.

31

Diagnosis

It had been twenty years since the accident on the glacier in Antarctica, and these new words stung me, but didn't surprise me. "You're suffering from post traumatic shock as a direct consequence of the accident," the doctor repeated. I simply stared at him disbelieving, unable to comment in my hurt. But I knew it was true.

It had been the first time I had considered seeking professional help for what was now becoming an increasingly serious problem. I was spiraling more and more into the depths of despair, and even I was noticing what was happening to me. Drinking had become an obsession as I fought the past trying to make sense of my flashbacks and I withdrew into my own intensely private world.

On entering the consulting room for the first time, the doctor had asked innocently and smiling, "And how can I help you?" He fixed me with his gaze as I struggled to answer his obvious and seemingly straightforward question. My answer took an age to come, but I managed eventually to blurt out the words that I needed help as my eyes filled and I stood up again and walked across the room, turning my back to him. I was aware that he turned his chair, keeping me in view, and he said nothing. He simply looked at me. He had seen his fair share of people suffering depression and guilt associated with some type of traumatic event that they had experienced, and his response, I was to learn later, had been well practiced.

As the clock ticked I became aware of overrunning my appointment time, but the doctor was not interested in the time, and as the silence continued he waited.

"I've been involved in an accident," I said, not really knowing where to start.

"Recently?" he said, but didn't appear taken aback when I retorted, "No—twenty odd years ago."

Not knowing what else to say, I fumbled. "And I need," I went on, "to stop these!" as I pulled a pack of Marlboro Reds, a sign of my recently restarted habit, from my black overcoat pocket, flashing the cigarettes embarrassingly in his direction. He laughed.

The doctor listened as we talked hesitatingly for nearly an hour while he carefully dragged something of the detail out of me. Pausing, backtracking, and never once losing eye contact, he began the long process of unburdening me of my grief and guilt until I was worn out, drained of all my already dwindling energy, and I felt empty. Crossing the waiting room with my head down to avoid any prying gazes from impatient people waiting for their turn to see the doctor, the thought that I had opened up and exposed myself to a complete and utter stranger, once again, hung in my chest, and the embarrassment hurt me. I felt utterly ashamed that I had betrayed my long dead friends by telling my innermost secrets.

It seemed easy for the doctor to say to me that it wasn't my fault. But what could a mere doctor in the confines of a cozy warm office some two decades after the incident know truly about what happened? He had only heard the story from me, and I expected to be judged and punished.

We met again a week later, the doctor and I. He had had time to consider my story and I had had time to recover from the initial shock of the revelation.

Our talking continued and he was able to help me take some tentative steps forward as he confirmed that this reaction of mine, so long after the event, was indeed quite normal. I was, in time, referred to a psychologist, who for the next few months picked away at my most intimate memories and background. In the course of that period of time, I was taken through hell, as I had to relive actions that I had not even thought about for such a long time. As I talked cold surged through my body and I would shiver in the warmth of the consulting room.

Memories, long buried, came flooding back and I descended further into my trough of depression as the old wounds were picked open and the blood and the puss oozed out—sometimes flowing freely, sometimes taking days after the consultation session to appear on the surface of my skin.

Following one particularly grueling session, I became lost and I could not see a future in anything. The memories were too painful, and I felt washed up by the fatigue and emotional turmoil I was being subjected to. I realized that if I was not gaining ground there was no point in carrying on, and thoughts of ending it all once and for all started permeating my every waking moment. Something had to change and it had to change fast.

Preparing to enter the psychologist's room for yet another consultation, my legs would simply not function. The door was a massive barrier that I could not break through and I turned on my heel and fled the scene. There was no way that I could speak; I was too confused and there was no energy left in me to fight it with. I needed to find answers and they were not coming to me.

My mountaineering days seemed far away in the past and still I felt no draw to the mountains. I had nothing to divert my thoughts into something more constructive as I wallowed in self-pity. But, with massive trepidation I forced myself to step onto the hillside once more in the middle of winter and gained the summit of a mountain after struggling through deep snow during one beautiful day. While all around me others were wearing their crampons and wind-proofs, I looked out of place and ill-equipped but felt a renewed energy as I raced from the top into the valley, remembering my love of the hills in less than a thousand feet of sliding, heel digging descent. I sat exhausted by my car in the snow at the end of the day and in the encroaching dimness I felt elated at my triumph. My lost mountains had been thrust back into my life and I had touched them again for the first time in years.

It seemed within weeks that I had overcome my problems and even remarked to a close friend that I was on top of the situation and happy, and I began spending less energy on thinking of the past and focussing on my guilt.

Months later on a yacht in the Caribbean I was waiting for the anniversary of the accident to come around yet again. The date was on my mind but I felt relaxed, rejuvenated, and fitter. The sun had browned me, the sea spray had weathered me, and I wanted to luxuriate in the water. I could swim once again, as I had the fitness to enjoy the effort. But with a bang the dangerous memories came flooding in once again as I toasted my long dead friends from a beach at the exact time of the accident. In my euphoria I thought I was home and dry, but it had simply been a holiday for my mind and reality once again stared me in the face.

Suddenly the falseness of my life hit me as I took the time to remember, not by choice, the scenes that had haunted me for so long. Once again I had totally lost control and I had no influence over when and where I would remember the ice and the snow and the horror. My self-confidence disappeared along with any self-esteem I had regained traveling down this blind alley, but now I had reached its end. This was not the way to deal with the problem; it was too ingrained in me and denial was not the answer. I still needed above all to find some answers and to understand.

32
Journeys End

In the close confines of the base and preparing for the rigors of winter life, I tried to fight on. The accident had happened. People had died. But it could not be the end. The base was going to be cut off from the outside world for at least a further six months, perhaps longer, and there was no way to escape from this world that I now viewed with fear and trepidation.

My role had been to travel and I could no longer face the prospect of miles of open spaces, on dangerous ground, on board a skidoo. I was unable to get over the events that had so scarred the landscape for me, and I began to hug more and more to the shirttails of the safe areas directly around the base.

Initially, I struggled badly and within a week after the accident I had let my frustration and temper go for the first time. I viewed everyone with suspicion. I had been in charge out there on the ice and two of my party had perished—I felt I was to blame as sure as if I had cut their ropes. Paranoia had begun to set in and I was simply not strong enough to resist it—no matter how many times the two members of the rescue party told me it had been simply a terrible accident, I did not trust their assessment. The Base Leader began to watch me closely and he followed me at times; he didn't need another death on his base and on his conscience.

Maurice and Simon appeared by my side more and more, talking, laughing trying to get me to relax with their gifts of alcohol and true friendship. They watched me, and they encouraged me to get through. I needed noise around me so

I made it myself when there wasn't anyone else to provide it, and everyone simply put up with the disturbance and contributed to my convalescence without criticism.

Nick disappeared into the sanctity and restrictions of his familiar workshop, emerging only at meal times and at the end of each working day. His fellow mechanic and him, two vastly different types, did not get along and they had arranged shifts where they did not have to meet during the working day. So Nick also continued in solitude during the long days.

At times I retreated into myself, thinking only of escape from the tight confines of the environment and of the ice and snow of the crevasse. The memories of the rope I thought was going to snap drove me towards madness during those early days, and I sought further help. The doctor gave me pills to sleep and after the first night I knew that if I could go one night without a pill I would have two for the next night, and deeper, less troubled rest. With three or four it would be even better, so I stayed awake and saved my pills and waited to have my own private soporific party in the confines of my bunk. On one occasion I took too many pills and I slept through the night, the next day, and the following night, eventually rising dulled by the drugs by mid-morning of that day. But I was happier because my exit was now that much nearer and my escape was that much the closer as more time had slipped by and I slept in the security of the hut.

I tried to help by encouraging as many people as possible to practice their self-rescue techniques on the ice cliffs and snow gullies around the base. We discussed the problems and I tried to show how difficult it could be in a crevasse with death trying to pull on your rope, but I did not force the issues strongly enough, and lacked confidence in my own abilities.

The mid-winter festivities, the highlight of the winter, came and went. A supreme effort was made to provide the base with a time to remember, and we all laughed and joked and talked for the full day. For weeks previously each of us had spent time enclosed in the secrecy of our own workshops creating our mid-winter presents. Each person made something and we would all draw lots and take that gift that had the number we had drawn. My contribution was a carved statue of a warrior with assegai and shield in dark wood. Not a statue of value, but emotionally I had put my soul into its creation and I lovingly spent hours filing and sanding the wood to create the image in three dimensions. Maurice jumped up as he pulled the number of my statue from the draw, and I felt enormously gratified that he appeared pleased.

But the gift I received was Simon's creation. A box in different woods and inlaid with the initials of the base and an outline of the Antarctic Peninsula. A cigar box, a box for letters, a box for jewels, I did not know what it was for, but in that moment it became my most prized and treasured possession. Years later the inlaid box was to become only one of a few items that did not find a place on the pyre that took many of my memories of Antarctica. It remains with me, and in pride of place, to this day.

Morale at the mid-winter festivities seemed high but there were strong undercurrents of violence and unhappiness. I knew nothing of the situation as I was largely locked in my own selfish world. One evening, some weeks after mid winter, I was finally let into the badly kept secret.

Seven of my dogs had been pulling me on a training sledge around the base during most of the day and I was encouraging them to get rid of their surplus energy, joining in with their obvious enjoyment. I was like a teenager promenading with his new car and showing off its sleek lines to any onlooker. The energy of the dogs was massive as they thrust the sledge forward loving the freedom of the run. A bump over a snowdrift and my right foot got caught between the foot brake and the sledge body, causing pain but nothing to worry about as I continued for a few more hours of escaping enjoyment.

That evening I had gone to my bunk reasonably early, forsaking the enjoyment of the formal Saturday after-dinner relaxation for the sanctity of bed. Within a couple of hours I rose once again to ease my aching foot and entered the bar area to the sight of bowed heads and silence. My questions fell on deaf ears until I was told that there had been a bit of trouble and someone had come off quite badly as the result of a beating. The unthinkable had happened and the shock acted like a painkiller on my aching foot.

Curtly, I was told that this was not the first time punches had been thrown or threats made to the worried and frightened base members. I had no idea there had been problems. In my own gloom and despondency, the unhappiness of others had simply passed me by. Before the ship had left for the return to UK months before at the end of the summer, some members who were due to spend the winter had considered returning also because of the threats of violence made to them by this one person. Only two or three of us had neither been hit nor threatened, and I instinctively knew that my turn would be next. Wilfred the ex-military man would not be accosted, as we all knew that to tangle with this quiet, intelligent man would be to enter a lion's den. Our Base Leader was above threats. So it had to be me, but when it came I was totally unprepared.

Weeks later the mood had subsided into one of quiet willing to get the winter over and done with as the unspoken undercurrent of threats remained. I had quietened somewhat and had regained some composure. My work had enabled me to find solitude and I was spending more and more time on my own in the sledging store, ambling on and contributing little of real value.

At a morning break I walked in the dim light to the main building and was greeted by three or four of the others drinking tea and laughing loudly. As I sat down with my cup, Maurice pulled my chair away and I landed on the floor on my back, but smugly with tea still in hand and little spilled. Looking up, I saw another of the base members race up the stairs. The first thing that struck me was that he was walking very quickly towards me and wearing his boots inside the building, working boots, steel capped and insulated, and in that instant I knew my time had come.

There was no time for me to get to my feet before his wildly swinging right foot missed my ducking head by a fraction. Amidst shouting and disorder, the red mist engulfed me as I administered my defense in the only way I knew how. No words, no pleading for calm, no backing away, and no mercy for my attacker. My pent up aggressions and sorrow overflowed onto my unfortunate assailant in a tirade of uncontrollable anger before any of the onlookers could intervene. It was all over in an instant of blood and bruising. Shocked, I stared at his battered face as he winced from my onslaught and I knew there would be no more violence.

Morse's response again showed his true leadership qualities as I informed him what had happened. Opening one eye, as he reclined in the lounge, he listened, sighed, looked at the chap next to him, and closed his eye again without a word. He would deal with it in his own time and although he appeared relaxed and almost disinterested, he was already formulating his plan.

My attacker, the attacker of so many of the personnel, was ill and he needed help. Forgetting his other aggressive acts I felt the confinement and the stress of the winter had worn him down and eventually it had snapped something inside him. This man, who was a good hardworking friend with such a fine sense of humor and a deep love of Antarctica, had succumbed and lashed out in his own personal grief and depression in the stress of seclusion.

It was an incident that caused me deep regret. I had put out a fire that had been raging in this man since the solitude of the winter began and which had remained unchecked for all those months. But in so doing I had hurt a friend when he had needed help, and in hurting him I hurt myself more. That he attacked no one else from that time gave me no solace.

We could not afford to have someone who was so ready to instigate violence in such close confines. It had all the etiquette of a typical public house confrontation, except we were not in a pub.

We were shackled to the interminable confinement of our civilization amongst an open wilderness of millions of square miles of snow and ice, and many of us were now hoping for an early release. Everyone kept their distance from the instigator of all the violence, and even in such tight confines it was days before he and I finally crossed paths. With a nod of understanding to each other, we both knew that the gap between us was territory that neither of us would invade again.

With the return of the sun, which due to bad weather conditions was obscured from view for days longer than it should have, spirits improved. We waited patiently outside on the day the sun should have reappeared after its trip to the north, all aware of the disappointing weather reports and knowing we would not be able to see and feel the relief we so needed and which we sought with a sight of the long lost sun. The next day at the appointed time the low thick cloud obscured our view again, as it did the next day and the next. But finally, even though we by now felt debilitated by nature's further taunting of our situation, the red streaks from over the horizon in the almost clear sky drew our admiring stares. Lights played across the sky from behind clouds and in front of the mountains. We watched intently as the sky turned deep red in the suns embarrassment of returning after its prolonged absence, before the heavens re-gained their confidence and retreated into greens and blues and orange and put on a show worthy of its power during the midday display. We stood amongst friends all equally captivated with cameras held loosely, unused by our sides in case we should miss anything of the spectacle.

With the return of the sun came renewed hope. We were nearly home, and I began looking forward now instead of constantly back. Antarctica was proving worthy of its name, dangerous in the extremes but drawing me into its arms once again.

With a start, early one late November morning, the dogs rose to their feet as one; they had heard it. The airplane engines, still far off, had come across the miles of clear uninterrupted space to the ears of the animals, and although we knew they were on their way we still only saw them a long time after the dogs knew they were coming. Greetings to the aircrew, new faces smiling after the ordeal of their long trip across half of the world in their flight of relief. The first flight out was going to have Wilfred and me aboard and I scrabbled amongst my gear to pack only what I needed.

Departures were hurried as Maurice, Simon, and I, standing on the snows of the airstrip, embraced our farewells, before I climbed aboard the noisy aircraft. Standing on the airstrip, my two friends looked up as I passed them overhead, then they looked down at their skidoos and mounted to head back to base, and they were gone.

A short flight and a treacherous landing on a steep slope leading to a mountaintop and the ship, my transport back to civilization, lay at anchor in the bay. An oasis of red amongst all this white.

New personnel on their way to the base stood in small groups at the edges of the runway. Orange clad and fresh, I guessed they had been told of the troubles as they kept their distance, or perhaps we told them to stay away with our unspoken words and body language. The big center back of the soccer game the year before, and the head Base Leader, greeted me and we talked briefly for only a few minutes. The accident was mentioned by me, but dismissed quickly, and Wilfred and I strolled the last remaining yards over the snow and through the screeching, interrupted Gentoo penguins to the waiting launch to be ferried out to the moored ship.

Within hours the engines were throbbing as the Antarctic began disappearing to our stern. Three of four days of seasickness later we were in the Falkland Islands. A flight to Buenos Aires and a further flight to Portugal and then home. A journey that had taken months on the way down took only days on return.

I sat amidst the noisy confines of the airliner over mid Atlantic and looked around at the few passengers aboard. The olive colored skins of the largely South American and Portuguese people, looking healthy and happy, contrasted deeply with the reflection I saw of myself in the mirror of the lavatory compartment. Drawn, weather beaten, straggly, unkempt long hair and beard, and dressed in a baggy green tartan shirt hanging loosely outside worn and tired jeans. The only reminders of the life I had just lead and where I had just come from. A place where I had spent the most harrowing, turbulent experiences of my life, not knowing at that time that I had not, nor ever would, shed the memories for as long as life itself.

33

Release: Twenty Years Later

At the very limits of my thoughts I felt there was something else, something I could grasp if only I could find the key. Talking and telling my story had helped to reduce my burden, and it had enabled me to take the first tentative steps forward. Visiting the church in Yorkshire had allowed me to confront one of the reasons why I had been so long with my suffering, and I had at last made my peace with Joe.

On a dismal day as the rain lashed against the window of my room, without thinking of what I was really doing, I began the torture of writing down my memories. At first the words seemed to flow in an unending stream, as the frustration of many years finally discovered release. I raced to get the words written in case they slipped back into the abyss where I had pulled them from and I became impatient with the day-to-day necessities of normal life. My computer screen faced me for long periods of every day as I dismissed anything from my life that could wait until my writing came to a natural end. But, in facing my fears, I found the answers I needed and I have provided my own version of those days; a story which could have perhaps helped me to understand years ago.

We were young and we thought we were immortal. There was nothing to fear when traveling across the wide spaces of the plateau and the glacier because we were young and the young view danger differently. But there is the reality of that wild world where a hole can swallow an unsuspecting traveler, as has happened countless times in the past. The expedition organizers employed me mainly because I had vast experience of the mountains and I knew I was safe in the terrain of the mountains. Crevasse rescue was as familiar to me as it is to any experienced mountaineer who spends much of their time on glaciers. I depended upon my skill as a mountaineer, trained in the mountains of the world and trained to travel on foot. In Antarctica, we taught ourselves to travel on board a machine that can travel as fast as a motorcycle while pulling the massive weight of a fully laden sledge. I always knew, without doubt, that if the skidoo I was traveling on were to fall into a crevasse, being tied to it would not save my life. And so it proved on the glacier with Ron, who was a highly experienced mountaineer, and Joe, who was not. I can offer no more proof than that; I saw the carnage. However, progress had dictated that the most efficient and effective means of travel was by skidoo. My life changed when, with inadequate rescue equipment that simply had not kept pace with the modern vehicles we used, I was unable to remove the skidoo from the crevasse to allow my friends to be brought to the surface.

Eventually, as my story unfolded on the page and as the memories surged back to the front of my mind, I began to question the role of the expedition organizers. Relatively untrained and ill-equipped people were sent into that environment. Those of us with some experience had to prepare those without any experience for every eventuality they would encounter in probably the most hostile continent in the world. How can people who have never experienced the threat of a real blizzard, when they are exhausted by the effects it creates, or the depth of a real crevasse hidden from view by a layer of snow, be expected to remain in control when these forces are threatening their life? In the aftermath of one of the most aggressive accidents ever to happen in that area, valuable experience was lost.

I wrote my report, I attended two non-technical, brief inquests in 1981 and 1982, but the next contact, which I had to instigate, with the expedition organizers was in 2000. By that time I was desperately in need of help, and in 2004 they were contacted again, having failed to respond to those earlier pleas. When eventually they agreed to meet me at their headquarters, they admitted to having no record of any earlier communication. During that meeting they witnessed the impact the accident had had on me, and was still having on me. They did stress verbally, that as far as they were aware, no blame had ever been attributed to any one individual.

The only technical point made during that meeting was that no person is now allowed to travel alongside a driver of a skidoo on the body of the vehicle. When this was stated I could feel the hairs on the back of my neck raise as I thought of Ron and Joe traveling together in that way as they broke through the bridge of the crevasse. But, I blurted out, that had been the rule all those years ago, and as proof of my adherence I had traveled with Joe on the back of my sledge all the way over the ice piedmont from the cape to the top of the glacier. Only when Joe changed position from the rear of my sledge to a position on Ron's skidoo was that rule ignored. I did not force my point further at the meeting, and it was only to be months later that I realized that I had probably been implicated in that mistake and that prior to that time, I had never stated my case in any depth. I had never had the opportunity and I had never been asked. In my so obviously traumatized state in the immediate aftermath, I had been incapable of emphasizing such a glaringly obvious point. As time went by, and as the expedition organizers failed to speak to me and listen to my opinions about the lead up to the accident, these answers became lost to me as they were thrust into the back of mind.

Joe had suffered intolerably in the face of the long lie up; he had lost his trust in me, as a direct result of the dangers we faced in the blizzard. When I asked him to change to Jon's sledge, high up on the side of the glacier, and as I was about to break trail, he did so willingly and with enthusiasm. He wanted to be close to his friend, an occurrence that takes place regularly when there is bad feeling or if there has been an argument. Inexperience of the hardship encountered meant he wanted to be as close as possible, and even though I objected to Ron vehemently about the position of Joe on the skidoo with Ron, Ron overruled me and I was too inexperienced to insist. That we should not have been in that area of the glacier is true. We tried but were allowed no margin for error and we paid the ultimate price for less than perfect navigation. On the flat of the plateau perhaps we would have escaped with Joe in that position on the skidoo, but amongst the crevasses we had little hope. It was inevitably their combined weight on that skidoo, a force pressing down in such a small area, which broke through the bridge of the crevasse and catapulted them to their deaths. I hold myself responsible for not insisting that Joe take his place on the rear of Ron's sledge, my only defense being I was overly preoccupied with the terrain and the difficulties of the predicament we were in. Once again, it can be confirmed that our level of traveling inexperience in Antarctica contributed to the deaths. Ron and I were mountaineers used to traveling on foot. I have had to live with the consequences of that massive mistake ever since.

With these written words has come a release; a release that has allowed me the time and solitude to concentrate on events that took place in a distant time in a distant land; a release that has allowed me to finally understand. In the claustrophobic confines of a small tent in a blizzard that wreaked havoc for days and nights, Joe, in his inexperience, suffered intolerably. That he was not prepared for such an arduous hell on earth is, for me, without question; I was with him, and it was upon me that he vented his displeasure. If we had made it back to base together I have no doubt that we would have argued and he would have told me in great detail of his fears and the hate he felt for me for putting him into that situation. I also know we would have traveled together again as our bond had become too strong. But that was not to be the ending we would have hoped for, and I have lived through the years in the knowledge that we did not have the opportunity to discuss our feelings with each other. Worse still, I did not have the opportunity to tell him that I was truly sorry for contributing to his discomfort and fear. He was not the first person to have felt the way he did as a result of being confined to a tent in such grossly difficult conditions.

Nick was also unprepared for the horrors of that blizzard and the reality of the problems we faced. But he faced everything thrown at us and came through with great strength of character and a determination I have not witnessed since. Nick could not understand why Ron and I reveled in that environment during the days of the storm. He said he could not understand why we would want to experience those places when he was willing to spend most of his time in the confines of the base. But he did understand our motivation in traveling and as a mechanic, he made a massive contribution in making sure we had the vehicles on which we could experience the environment.

Ron was always aware of the dangers and the problems. He was a brilliant mountaineer who loved that environment with a passion. In the tent in the blizzard, he had been relaxed and confident in the face of extreme danger. During one of our conversations, he had spoken of an environment so beautiful it was beyond true description or the ability of film to truly record its beauty. Some of his last words to me confirmed that feeling. Ron knew his strengths and abilities better than anyone.

As for me, during the enforced lie up as the blizzard was so aggressively taunting us, I had to simply accept the problems we encountered were part of the dangerous yet beautiful world that is Antarctica. The past has poured through my veins and I have learned to face my shadowy corners in the full light of day. With the support of true friends who accepted and understood all my fears and guilt and who encouraged me to bare my soul and be true to myself, I have moved on. Those friends chided me. "Face the past and acknowledge the difficulties, and the mistakes, but more importantly rid yourself of guilt and live your life. None of us have more than this life and it seems a waste to squander what little time we have."

I have lived with the legacy of Antarctica for more than twenty years now. One short day in that continent, in a land that I knew so briefly, pulled and distorted my life into the turmoil of total desolation. Now I have made my peace. My path forward is plain to me. I remember those friends who died on that glacier. But I remember them in life.

The End

Printed in the United Kingdom by
Lightning Source UK Ltd., Milton Keynes
142467UK00001B/20/P